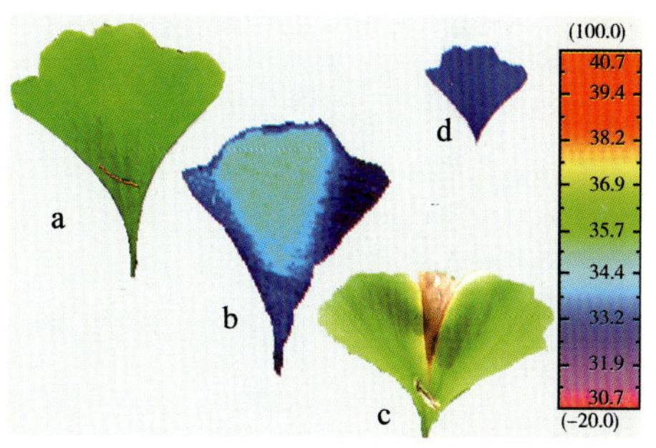

图 1 切脉时银杏幼叶的 RGB 图像(a)、切脉 6 天后的 RGB 图像(c)以及该叶切脉前(d)/后(b)的热红外图像。叶片切脉的部位用“—”标记

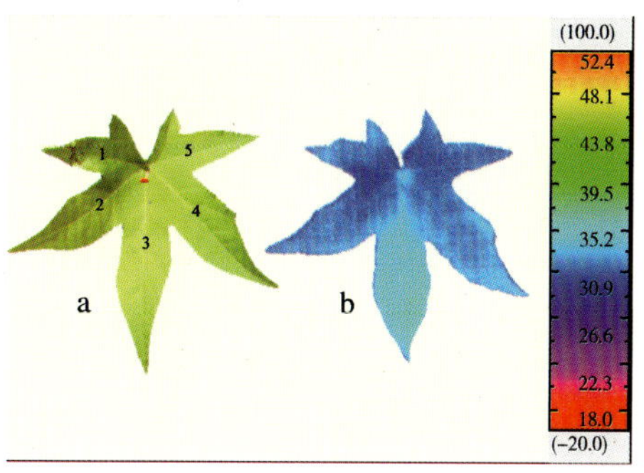

图 2 仅第三主脉被切断的北美枫香叶的 RGB 图像(a)和热红外图像(b)。

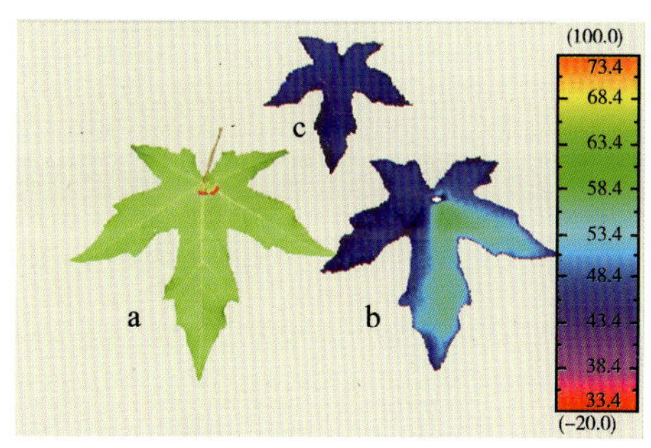

图 3 第三和第四主脉被切断的北美枫香叶的 RGB 图像(a)、热红外图像(b)和该叶切脉前的热红外图像(c)。

图 4 第二、三和四主脉被切断后的北美枫香叶的 RGB 图像(a)和热红外图像(b)。

图 5 北美枫香叶主脉被切断后经历持续的胁迫后导致叶变色的典型特征:包括第二、第三和第四主脉被切断的(a),第三和第四主脉被切断的(b)和只有第三主脉被切断的(c)。

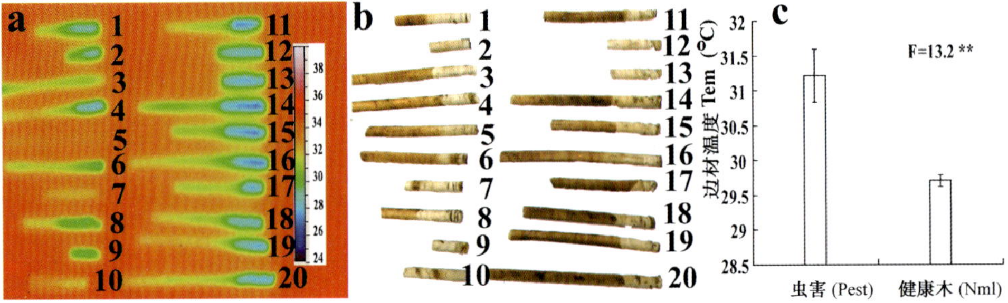

图 6　柏肤小蠹和双条杉天牛侵袭的侧柏主干生长锥芯之热像、数字图像和边材热像温度的对比：a. 受害木（左 1～10）和正常对照木（右 11～20）的热红外图像；b. 图像 a 中所对应的数字图像；c. 受害木（Pest）和正常木（Nml）边材温度（Tem）对比柱状图。

Fig. 7　RGB and thermal images of the front and back surfaces of a Japanese spindle leaf as the temperature decreased after hand heating. a. Front surface of a leaf with a clear tip and a margin scorch. b. Back surface of the leaf in (a) shows the thinner major or minor vein in the tip area. c. Thermal image of the leaf in (a) with almost the same low temperature at the tip and margin areas. d. A leaf with significant water trance at the base of the leaf and along the major vein. e. Thermal image of the leaf in (d) as the temperature decreased after hand heating. f. Thermal image of a Japanese spindle leaf, more than half of which was crosscut.

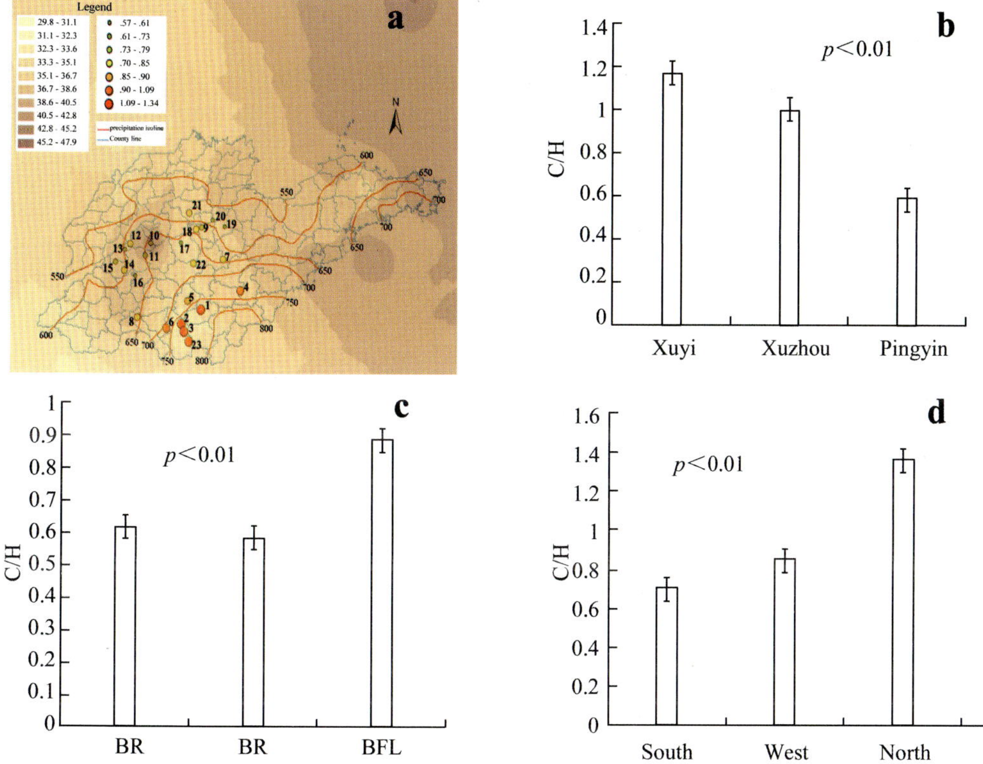

Fig. 8 Crown shape variations of *Platycladus orientalis* trees in different environments: a. the Arcgis map shows the Max30twp index, crown shape index (CSI) and precipitation isoline during the period over 0 ℃ (—) of Shandong Province, China; b. crown shape index (C/H) for Xuyi, Xuzhou and Pingyin; c. crown shape index (C/H) for the trees at sites with different soil types, in which BR is brown rhogosol and BFL is brown forest loam; d. crown shape index (C/H) for the trees at different slopes on the same hill.

Fig. 9 A persistent reddening juvenile sweetgum leaf with five major veins signed with Ⅰ, Ⅱ, Ⅲ, Ⅳ and Ⅴ, respectively (a, PM and VS on Jun. 1). It was severed at the base of Ⅱ, Ⅲ and Ⅳ major veins with mark "—" and divided into ten half-lobes in the order against hour hand. The same leaf (b, PM on Aug. 1) with the persistent reddening area at stressed part (S, half-lobes of 4, 5, 6 and 7), greened area at non-severed part (N, half-lobes of 1, 2, 9 and 10) and the transitional area between them (T, half-lobes of 3 and 8). Another persistent reddening stem-developed juvenile leaf (c, VS on Jun. 1 and PM on Aug. 1) and its thermograph (d, PM on Aug. 18). The persistent reddening test had been duplicated more than six leaves and these two were typical examples.

Fig. 10 A normal chlorophyll mature leaf (a, PM and VS on Aug. 9) and the stressed area advanced reddening for the same leaf after Ⅱ, Ⅲ and Ⅳ major veins were severed (b, PM on Sep. 21) with the severed positions marked with "—" and water content (W) at non-severed area (N), transitional area (T) and stressed area (S). A slightly advanced reddening leaf at the stressed area (c-S) after Ⅲ and Ⅳ major veins were severed (c, VS on Jul. 6th and PM on Sep. 6th), in which the number next to "L" is the leaf stomata conductance value and the number next to "G" is the ratio between green and luminance values from RGB image. The thermograph (d, VS on Aug. 11st and PM 18th minutes after VS) of a leaf whose were Ⅲ and Ⅳ major veins were severed with different image temperatures (t). The thermograph (f, PM on Aug. 18th) and the RGB image (e, VS on Jun. 1 and PM on Aug. 18th) of a shade leaf. An entire red leaf (g, VS on Aug. 9th and PM on Sep. 21th) with advanced reddening area and an entire green leaf (h, VS on Jul. 5th and PM on Sep. 4th) without advanced reddening area. Ⅱ, Ⅲ and Ⅳ major veins of leaves in e, f, g and hare h were severed. The advanced reddening test had been duplicated more than twenty times and here are some typical examples.

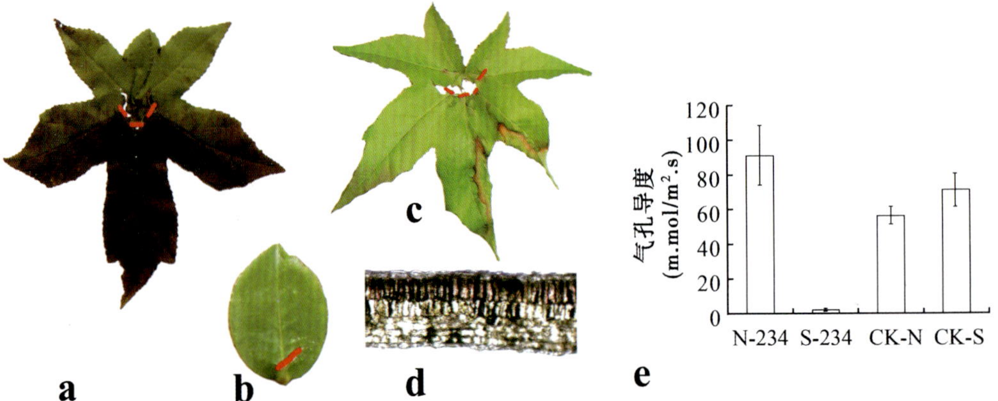

图 11 一片 2、3、4 主脉基部切断后表现出局部变红的北美枫香绿色成熟叶(a);一片无显著变化的山茶叶(b);一片切断了 2、3、4、5 主脉的北美枫香嫩叶,在第 3 和第 4 裂片上出现了明显的焦枯症状(c);北美枫香变为红或紫色之叶片的横切面显微图像(d);一片 2、3、4 主脉基部切断的北美枫香叶绿色区域之气孔导度(e,N-234)和红色区域的气孔导度(e,S-234),以及未切断主脉的北美枫香叶的对应部位之气孔导度(e,CK-N 和 CK-S),气孔导度测定重复 3 次。

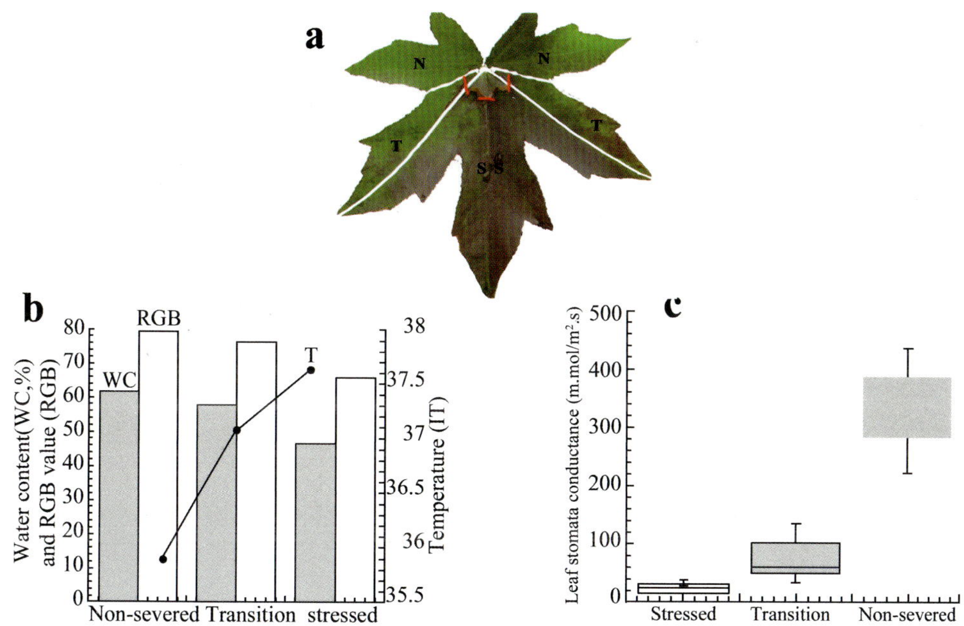

图 12　主脉切断之北美枫香叶片的水分胁迫和变红特征。图 a 为主脉基部切断的北美枫香叶的局部变红，其中包括未切脉区（N）、过渡区（T）和胁迫区（S）；叶脉切断部位由“—”表示。图 b 为未切脉区、过渡区和胁迫区的水分含量（WC）、热像温度（IT）及 G/L 值（$n=3$）。图 c 为未切脉区、过渡区和胁迫区的气孔导度值（$n=10$）。

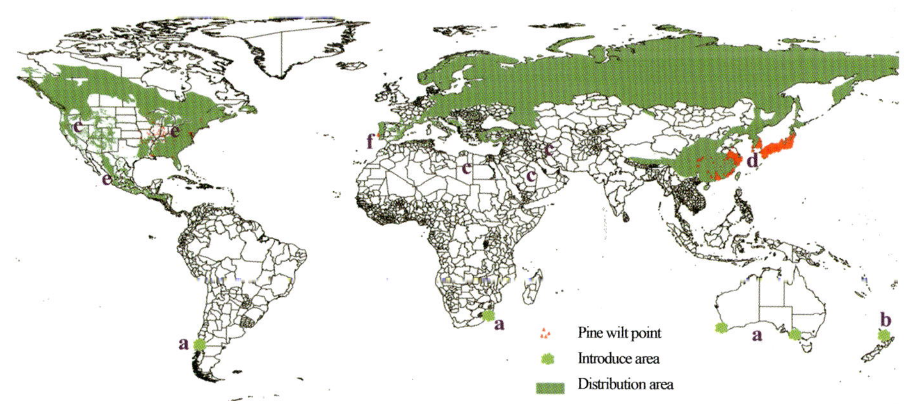

图 13　a 为世界松树分布和引种栽培地区轮廓图和松材线虫发生地域图；引种栽培松树的澳大利亚、智利、南非；b 为新西兰松树栽培区；c 为北纬 25～35 度没有松树自然分布的地区；d 为日本、中国、韩国松树枯萎病流行区；e 为美国和墨西哥松树枯萎病局部发生区；f 为葡萄牙地中海气候的松树枯萎病发生区。

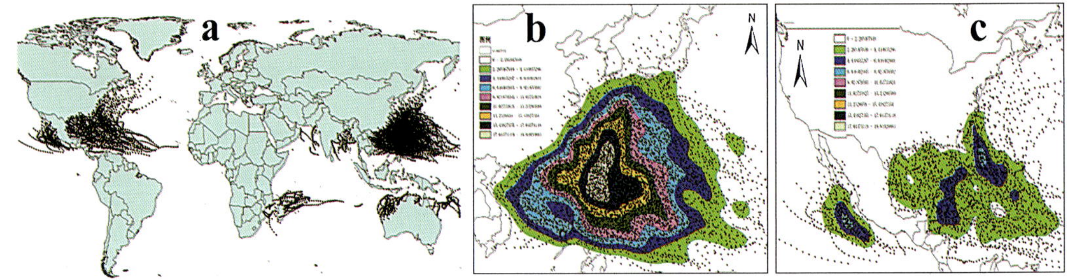

图 14　图 a 为世界范围内最大阵风风速超过 113 节的超强台风/飓风(每日 0、6、12、18 时中风速大于等于 64 节的时段)的点分布图;图 b 为西北太平洋区域该类数据的 ARCGIS 9.3 系统 Kernel 点密度分布图;图 c 为北大西洋区域此类数据的 ARCGIS 9.3 系统 Kernel 点密度分布图。

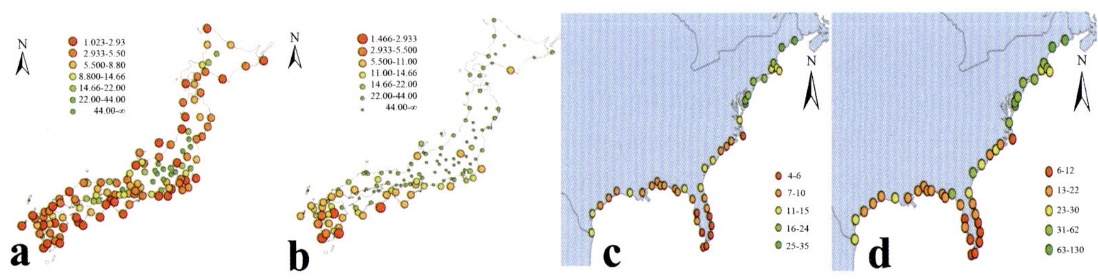

图 15　日本各地和美国大西洋地域最大阵风风速超过 33 m/s(一级)和超过 42 m/s(二级)的台风(飓风)回归年值的比较。图 a 为最大阵风风速超过 42 m/s 的台风回归年值;图 b 为最大阵风风速超过 33 m/s 的回归年值;图 c 为 1 级以上飓风回归年;图 d 为 2 级以上飓风回归年。

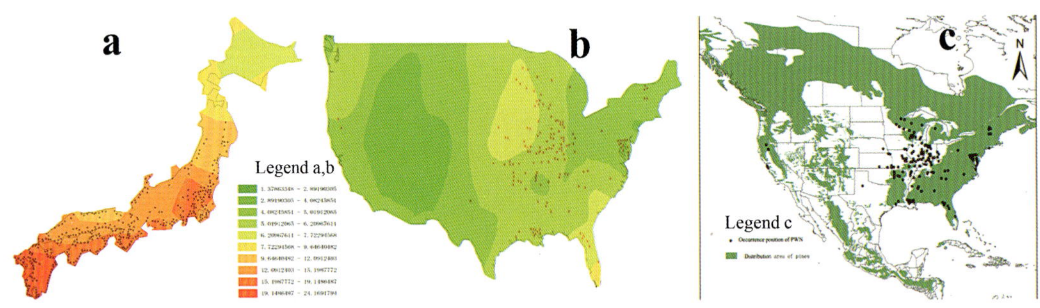

图 16　图 a、图 b 分别为用日本和美国各县/州的降水波动指数通过 ARCGIS 9.3 系统的径向基函数进行的地统计分析结果图;图 c 为北美地区松树分布或潜在分布特征图及枯萎松树中发现松材线虫的样点。

树木生态信息学论文精选

王 斐 等著

山东大学出版社

图书在版编目(CIP)数据

树木生态信息学论文精选/王斐等著. 一济南:
山东大学出版社,2018.8
ISBN 978-7-5607-6117-6

Ⅰ.①树… Ⅱ.①王… Ⅲ.①树木学－植物生态学－信息学－文集 Ⅳ.①S718.45-53

中国版本图书馆 CIP 数据核字(2018)第 185764 号

责任编辑:李昭辉 郑琳琳
封面设计:牛 钧

出版发行:山东大学出版社
社 址 山东省济南市山大南路 20 号
邮 编 250100
电 话 市场部(0531)88363008
经 销:新华书店
印 刷:济南万方盛景印刷有限公司
规 格:787 毫米×1092 毫米 1/16
13.75 印张 327 千字
版 次:2018 年 8 月第 1 版
印 次:2018 年 8 月第 1 次印刷
定 价:36.00 元

序　言

进入21世纪以来，世界逐步进入了信息化时代。物联网、大数据的应用使当今科学研究领域发生了重大变化。作为一门新兴的学科，生态信息学正在不断显现出突出的优势和潜力。本书作者在2006～2009年曾就读于日本鸟取大学，期间参与过日本文部省“科学研究费补助金特别研究项目”。2009年11月归国后，先后获得过国家自然科学基金项目(31170671)、山东省科学技术发展计划项目(2012GNC11107)、国家人力资源和社会保障部归国留学人员择优资助项目(20110026)的资助。在此期间，本书作者应用数字图像解析法、热红外成像等生态信息技术，通过对树木表型特征的量化解析，深入探讨了树木对极端环境事件的响应特征及机理。此外，还用生态信息学手段致力于树木生态生理学、灾害气象学、数字图像与光谱分析以及灾害评估等方面的研究。本文集收录了作者2008年以来公开发表的科研论文，鉴于时间紧迫，故没有把英文文献翻译成中文。作者认为，或许这样更能体现论文的原义。

作者在相关领域的研究中，应用切断叶片主脉等方法研究了许多树种在水分失衡状态下的能量平衡及响应特征。结合对叶片气孔导度、光和作用、光化学特征、花色素苷、超氧化物歧化酶、苯丙氨酸解氨酶等的观测(尤其是饱和蔗糖渗透试验)，对植物系统性响应症状的出现和机理进行了探索，提出了叶片“Λ”“V”症状的识别法则和途径，为简便而快速地区分气象灾害和一些原生或次生病害症状奠定了基础。与此同时，在松针的色素提取过程中，作者发现了绿色松针在浸提脱水过程中花色素苷形成的过程，为以后分析松树枯萎的机制创造了条件。

相关研究继承和发展了“蒸腾表面积缩减”这一理论学说，深入研究了树叶系统性受害的特征，对研究树木响应极端干热环境的机理具有非常重要的意义。作者还用热红外成像技术直观地再现了一些树木断脉叶片、自然胁迫叶片以及渗透胁迫试验枝叶蒸腾表面积缩减的渐进过程；依据水分和能量平衡理论，从结构功能统一理论出发，再现了蒸腾冷却失调和水分、能量失衡下的叶色早红和续红；依据全球气候变化中极端气象事件频发和相关现实，结合使用数字图像解析方法，运用输导系统的功能转换理论，合理地解释了一系列难以理解的自然现象和棘手的问题，如松柏的整株枯萎等。

对一些植物疑难杂症的研究发现，植物病害发生之前，许多植株往往因种种不良环境的影响而出现衰退，从而进一步增加了问题的复杂性。环境因素在生物病害的发生过程中可能起次要的，但却是关键的激发作用。有研究表明，在一些乔木、灌木病害的处置过程中，当解决了限制性环境因素的生理胁迫之后，一些侵染性病害也就迎刃而解了。

植物或树木有机体的生命过程是复杂的，这种复杂性远远超过有机体周围环境的复杂性。它不仅受制于土壤、水分、光照、养分等环境因素，而且还依赖于内在的遗传品质和生长状态，甚至受多因素的综合作用和影响。生态环境因素之间存在相互作用，而且一个因素不足时会由另一因素来补足。种种事实证明，生命的和谐在于结构和功能的统一，在于遗传和环境的协调，在于内因和外因的平衡。

王　斐

2018 年 5 月 5 日于济南

目　录

第一部分　生态信息

用红外热像法检测一些树木枝叶温度的研究*

王　斐[1]　山本晴彦[2]

(1.山东省林业科学研究院,济南,250014;2.日本山口大学农学部,日本山口县山口市,753-8515)

1.引言

红外热成像是一种用来探测目标物体的热红外辐射,并通过光电转换、电信号处理等手段将目标物体的温度分布图像转换成视频图像的技术。植物对 8～14 μm 波段辐射光谱的高吸收率和发射率(发射系数 0.97～0.99)表明,该波段的热红外光谱与植物温度有关[1]。植物或树木的温度值是其能量代谢的结果,温度的高低影响其辐射红外光谱的数量和质量。在自然环境中,它们常会遇到接受过量光能和热能的事件,而光保护机制又会把多余的光能转换成热量[2]。植物组织通过三种主要方式进行散热,即长波辐射、热对流和水分蒸腾。蒸腾冷却作为一个生理学概念已有很久的历史[3]。水分在蒸腾过程中可吸收蒸发潜热,这使其成了植物或树木最有效的自然散热方式[4]。较低的深层土壤温度和树冠遮掩下的低土温本身也使得蒸腾冷却更加有效[1]。水是生命的基础,其巨大的比热值而带来的冷却作用使众多树木的温度变幅小于大气温度。所以,曾有人认为在有足够的水分供应的条件下,过量的光照和高温不足以对已经适应了环境的植物造成伤害[5]。在水分亏缺的条件下,常可观测到树木大于 40 ℃的高叶温[3]。尽管叶温受很多外界条件的影响,但它常被用来反馈植物水分状态[6]、气孔的关闭[7～9]和蒸腾的衰减[10～12]。有研究表明,监测气孔导度比监测水势能更好地反映植物对土壤干旱的响应[7]。鉴于常用的"气室法"在测定气孔导度时只能获得点状数据,基于水分蒸散冷却、气孔关闭使叶片升温的原理,热红外成像法是一种有前途的选择[10]。在极端水分亏缺的环境条件下,树木往往会因过度失水而出现枝叶枯死,业已枯

* 基金资助:日本文部省科学研究费补助金特别研究促进费。通讯作者:王斐,1959 年生,日本鸟取大学博士,山东省林业科学研究院研究员。

死的枝叶含水率呈阈值曲线式变低。这成为应用热红外成像技术鉴别枯死枝叶的另一依据，因为热像是研究植物的空间和时间异质性特征的有效工具之一[13]。

红外热像法已成功应用于实验室内测试影响气孔导度的胁迫，而在室外的应用并不顺利，常因环境条件的不同而波动变化[14]。在室外热像拍摄的过程中，光照和遮阴的枝叶之间存在较大的温差。一些研究人员喜欢光照条件下的测定，也有些人认为在遮阴状态下测定更好[15]。尼尔森[16]曾提示，温度变化的动态过程对于识别不同程度的生物或非生物胁迫更加重要。事实上，在热红外图像的检测中，主动加温技术已在农林业以外的一些领域中得以应用[17]。室外测定时的各种干扰也使不同热像之间缺乏可比性。为了增加可比性，一些研究人员曾将各种参照模型应用于植物热像测温[15]。在本研究中，阳光直射的环境和专门设计的断脉导致北美枫香(*Liquidambar styraciflua L.*)和银杏(*Ginkgo biloba L.*)叶片局部水分亏缺和增温，这使其成了研究叶温和蒸腾冷却变化的特殊材料。在断脉叶片上温度梯度的存在使得热像拍摄更加容易，系统误差也较小。在同一叶片上断脉和未断脉裂片的可比性远远高于不同热像间的绝对温度值的可比性。在此基础上，在室内外用红外热像法检测了水分亏缺状态下一些树木枝叶的温度。

2. 材料与方法

本研究在日本山口大学农学部进行，研究材料来自山口大学校园内若干北美枫香根蘖植株和几株街路树、部分枝叶枯死的新移植四照花(*Cornus kousa Buerg.*)植株、银杏防护林带内的银杏树以及红叶石楠(*Photinia glabra Thunb.*)绿篱的个别植株。所观测的景观树之共同特点在于其树体均不高大，便于枝叶采集和热红外图像的拍摄。

银杏和北美枫香叶脉总是从叶基切断(见图 1 至图 5)，叶片切脉的部位用“—”标记。银杏叶典型的二叉叶脉和北美枫香典型的掌状叶脉使得它们对切断叶脉比较敏感。本研究正是利用这一特点，专门设计了不同切脉方式，诱导出叶片内部的水分胁迫梯度和高温区域(见图 2 至图 4)。试验选择平展的银杏幼叶，在叶片的基部斜向切断大部分叶脉(见图 1a)。北美枫香具有典型的掌状叶脉，有 5～7 个裂片，每个裂片上有一条主脉(见图 2a)。为了便于比较，全部选择平展的 5 裂叶用于切脉试验，裂片和叶脉均按逆时针顺序编号排序(见图 2a)。试验设计了 3 种叶脉切断类型，分别为切断 1 个叶脉(第三主脉，见图 2a)、2 个叶脉(第三、第四主脉，见图 3a)和 3 个叶脉(第二、第三、第四主脉，见图 4a)。银杏和北美枫香叶片的热红外图像均是在实地活体拍摄的。四照花枯梢枝条的热红外图像拍自离体的枝段(见图 6b)。该枝段从始枯节开始向活着和死亡的部位各截取 3～4 cm，且保留了从该节点萌生的新叶。为了观察红叶石楠叶痕处的蒸腾冷却现象，选择在阳光直射下的枝条，摘掉从顶端起第三或第四叶，使白色叶痕暴露在阳光之下，然后活体拍摄红外热像和 RGB 图像。

红外热像用 NEC TH7100 热红外相机(波长 8～14 μm)拍摄。该相机的测温范围为 −20～100 ℃，最小感温能力 0.06 ℃。测定时手持摄像机于目标叶片上方 50 cm 处，调焦到清晰。红外热像是在上午 9:00～12:00 的自然阳光照射加温过程中拍得的。选择平展的叶片，先拍摄一张未切脉的对照热像，切断叶片主脉后连续拍摄红外热像。选择不同部位间温差最大的典型枝叶进行结果分析。叶片或枝条的抽出是用Photoshop软件的磁性套索工具完成的。

RGB图像用Canon CCD数码相机(IXY6.0)拍得,然后同样用Photoshop软件的磁性套索工具分离出叶片或枝条。

3. 结果与分析

3.1 切脉后银杏幼叶的叶温升高和局部枯死

独特的二叉叶脉直接终结在叶缘且叶脉之间几乎没有吻合,这使银杏叶对切脉、机械损伤和有害生物的危害非常敏感[18]。切脉后的银杏叶(见图1a)不久就表现出在切口的后方形成明显的放射状高温区(见图1b),其热红外图像明显不同于切脉前叶片的热红外图像(见图1d)。切脉后切口后方的局部叶片处于严重的水分亏缺状态而出现了明显的局部萎蔫,持续的萎蔫使得远离未切断叶脉的部位发生局部枯死(见图1c)。从图1b和1c的比较可以看出,枯死部位的相对面积明显小于高温部位的面积,这应该是来自于未断叶脉部位的侧方水分供应所致。叶片发生局部的枯死并不妨碍未断叶脉区域的正常生长,从而使叶片分化成典型的扇形叶(见图1c)。

图1 切脉时银杏幼叶的RGB图像(a)、切脉6天后的RGB图像(c)以及该叶切脉前(d)/后(b)的热红外图像。叶片切脉的部位用"—"标记(参见彩色附图1)。

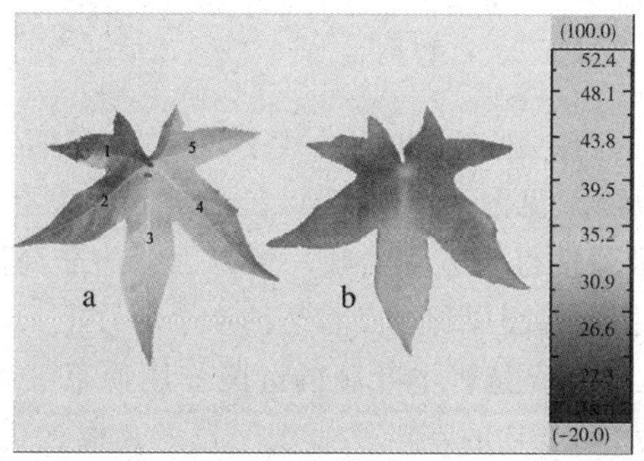

图2 仅第三主脉被切断的北美枫香叶的RGB图像(a)和热红外图像(b)(参见彩色附图2)。

3.2 北美枫香叶切脉后局部的增温和变红

观察发现，北美枫香典型的掌状叶脉和仅裂片间的细脉存在吻合的特点使它们对切脉也较敏感。对断脉后的红外热像的观察表明，在断脉裂片上出现了明显的高温区域（见图 2b、图 3b、图 4b），且同一叶片在切脉前后有明显的温度差别（见图 3b、图 3c）。然而，断脉部位的不同高温区（见图 2a、图 3a、图 4a）出现的位置不一（见图 2b、图 3b、图 4b）。第三主脉被切断的叶片只在第三裂片上呈现轻度地增温，且距离水源越远温度越高（见图 2b）。而在第三和第四主脉被切断的叶片上第三裂片的右半侧和第四裂片的左半侧出现了明显的高温区（见图 3b）。在第二、第三和第四主脉被切断的叶片上，高温区则出现在第二裂片的右侧、第四裂片的左侧和整个第三裂片（见图 4b）。

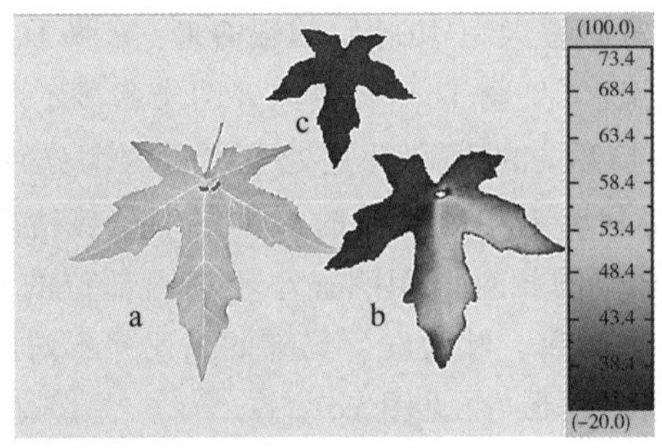

图 3　第三和第四主脉被切断的北美枫香叶的 RGB 图像（a）、热红外图像（b）和该叶切脉前的热红外图像（c）（参见彩色附图 3）。

图 4　第二、三和四主脉被切断后的北美枫香叶的 RGB 图像（a）和热红外图像（b）（参见彩色附图 4）。

显然，在第二、第三和第四主脉被切断的叶片上，第二和第四主脉似乎成了妨碍水分侧向输导到远离水源一侧的屏障（见图 4b）。然而，在正常的叶片上没有这样的现象发生（见图 3c）。正是这种主脉的屏障使得切脉后的叶片可以区分为非高温区和高温区（见图 4b）。非高温区显然位于主脉未切断的裂片和过渡状态的半个裂片上。高温区则位于主脉屏障的另一侧，即远离水源的部位。正是这种主叶脉的屏障作用使第三和第四主脉被切断的叶片只在第三裂片的右半侧和第四裂片的左半侧呈现明显的高温区（见图 3b）。在切断第三主脉的叶片上高温部位仅出现在该裂片的尖端，显然是因为没有明显的主脉屏障存在的缘故（见图 2b）。据观察，在最极端的阳光直射条件下，第二、第三和第四主脉被切断的叶片在高温区的尖端甚至会在1小时之内焦枯(图4a)。在第三和第四主脉切断的叶片上,焦枯症状往往始发于第三和第四裂片的结合部,这与叶片温度的增加趋势一致,意味着主脉切断后冷却水的供应终止或减少。在这一区域过量的光能和高温激发了叶片的保护反应,并分离出严重胁迫的部位以削减蒸腾表面积[19](见图4a,第三裂片尖端)。植物通过叶片表面积的减少而实现叶片水分平衡,并维持叶片其余部分必要的蒸腾冷却,尤其是正在快速伸展的幼叶。

图5 北美枫香叶主脉被切断后经历持续的胁迫后导致叶变色的典型特征:包括第二、第三和第四主脉被切断的(a),第三和第四主脉被切断的(b)和只有第三主脉被切断的(c)(参见彩色附图5)。

在适中的环境条件下,在叶脉切断一段时期后,敏感叶的高温区常出现变红或失绿的光保护现象(见图5a、图5b、图5c)。在第二、第三和第四主脉被切断的叶片上,典型的红色区域出现在第二裂片的右半区、第四裂片的左半区和整个第三裂片(见图5a)。在第三和第四主脉被切断的叶片上,典型的红色区域出现在第三裂片的右半区和第四裂片的左半区(见图5b)。鉴于仅第三主脉被切断的叶片没有明显的侧向主脉输水屏障,故只在第三裂片上看到不很显眼的失绿现象(见图5c)。据观察,变红的叶片很少发生叶片焦枯,这意味着红色保护层减少了光能的接收。非高温区域常保持绿色,且随时间的推移高温区和非高温区的颜色差异更加明显。

图6 四照花枝条枯死部位(a,节之下)和存活部位(a,节之上)的RGB图像和枝条枯死部位(b,节之下)和存活部位(b,节之上)的热红外图像。

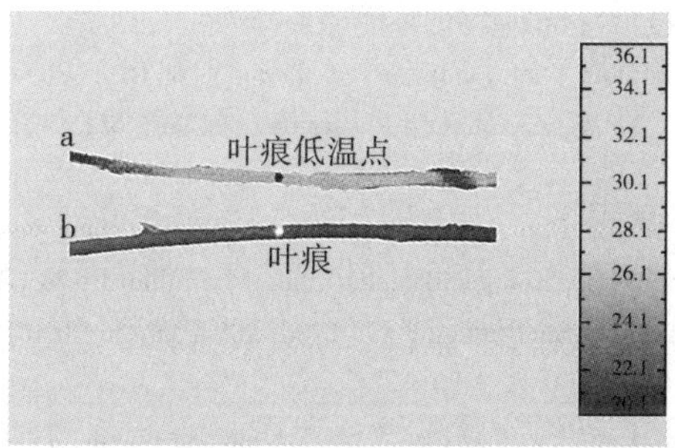

图7 红叶石楠叶痕的热红外图像(a)和RGB图像(b)。

3.3 四照花枯梢的高温和红叶石楠叶痕的低温

据观察，在极端干热的环境胁迫下四照花移植植株不仅会出现叶片从叶尖到基部逐渐枯死的情况，而且会发生枯梢现象。枝条的枯死往往截止于节点部位。枯梢部位的含水率很低，一般在10%左右，在极端干旱的环境中不足5%，从而使枯梢枝条的含水率从基部到顶部的变化呈现反逻辑斯蒂曲线趋势，即阈值“关闭开关”型趋势。水分含量的这种阈值变化趋势以节点为转折点，可鲜明地表现在热红外图像之中（见图6b），尽管在RGB图像中并不明显（见图6a）。除此之外，还可以看到明显的切口低温效应，因为有水分不断从切口处蒸腾出来（见图6b，上部切口）。这种切口效应还可以在摘掉叶片后的红叶石楠叶痕（见图7b）处看到，因为在红叶石楠枝条的热红外图像上可以看到非常显眼的叶痕低温点（见图7a），这种低温点随时间的推移而消失，表明蒸腾冷却使得枝条的局部温度低于没有明显蒸腾的部位。

4. 讨论

银杏和北美枫香等树种的叶片通常会在切断叶脉后不久就因水分亏缺和冷却水分不足而导致叶温升高，尤其是在阳光直射或暑热天气条件下。在这些环境中，用红外热像法可探测到由于成倍增加的水分不平衡而致的高温区。叶片变红区域与高温区的吻合表明蒸腾冷却的不足诱发了北美枫香叶光保护机制的启动，从而导致局部叶变色。本研究正是北美枫香叶在主脉被切断的条件下，通过监测阳光照射升温的过程，用红外热像检测蒸腾衰减的实例。在干旱少雨、夏季炎热的地区由于蒸腾降温功能的失调而使一些树种易发生叶尖叶缘枯死和抽梢死亡。作为一种探测物体热红外辐射光谱变化的工具，用热红外成像技术监测枝叶温度的变化应该有益于叶尖叶缘枯死和抽梢死亡以及蒸腾衰减的判别和确定。

主要参考文献

[1] Rosenberg N. J. Microclimate: The Biological Environment[M]. New York; London: Jone Wiley & Sons, Inc. ,1974.

[2] Donald R. Plant Physiol[J]. Springer, 2001,125:29.

[3] Gates D. M. Transpiration and Leaf Temperature. Annu. [J]. Rev. Plant Physiol, 1968, 19: 211-238.

[4] Fitter A. H. ,Hay R. K. M. Environmental Physiology of Plant[M]. San Diego, Landon: Academic Press, 2002, 162-190.

[5] Mansfield T. A. ,Jones M. B. Photosynthesis: Leaf and Whole Plant Aspects. In Hall M. A. eds. Plant Structure,Function and Adaptation[M]. Landon, Basingstroke: Macmillan Press LTD. , 1976, 315-316.

[6] Jones H. G. Application of Thermal Imaging and Infrared Sensing in Plant Physiology and Ecophysiology[J]. Adv. Bot. Res. 2004, 41: 107-163.

[7] Jones H. G. The Plant Cell Plant Direct The Arabidopsis Book[J]. Plant Cell Environ, 1999, 22: 1043-1055.

[8] Chaerle L., Caeneghem W. V., Messens E., et al. Trends in Plant Virus Epidemiology: Opportunities From New or Improved Technologies[J]. Nat. Biotechnol., 1999, 17: 813-816.

[9] Chaerle L., Van Der Straeten D. Trends in Plant Science, 2000, 5(11): 495-501.

[10] Grant O. M., Chaves M. M., Jones H. G. Physiologia Plantarum. 2006, 127: 507-518.

[11] Jones H. G., Stoll M., Santos T., et al. Evaluation of Grapevine Water Status from Trunk Diameter Variations[J]. Exp. Bot., 2002, 53: 2249-2260.

[12] Prytz G., Futsaether C. M., Johnsson A. New Phytol. 2003, 158: 249-258.

[13] Chaerle L., Leinonen I., Jones H. G., et al. J. of Exp. Bo., 2007, 58(4): 773-784.

[14] Grant O. M., Tronina L., Ones H. G., et al. J. of Exp. Bo. 2007, 58: 815-825.

[15] Jones H. G., Leinonen L. J. Agric. Meteorol., 2003, 59(3): 205-217.

[16] Nilsson H. E. Can. J. Plant Phathol. 1995, 17: 154-166.

[17] Yang X. L., Du L. L., Feng L. C. Laser and Infarred, 2007, 37(11): 1188-1191.

[18] Shull C. A. Plant Physiology, 1934, 9: 387-389.

[19] Wang F., Yamamoto H., Ibaraki Y. J. Forestry Research., 2009, 20(3): 254-260.

(原文发表于《光谱学与光谱分析》2010, 30(11): 2914-2918.)

侧柏衰弱木和蛀干害虫受害木的热红外成像检测*

王　斐[1]　吴德军[1]　翟国峰[2]　臧丽鹏[3]

(1. 山东省林业科学研究院,250014,济南; 2. 山东省沂源县亳山林场;3. 山东农业大学林学系)

1. 引言

历史上,研究林木和森林健康的相关文献时有报道,韦林[1]曾在1980年提出,可将叶面积与早材面积之比作为林木活力指数。另外,也有众多指标体系被用来衡量森林树木的健康状态[2,3]。有学者利用树干液流观测设备研究了许多树种树液流动的特性及其水分代谢特征,其中也包括侧柏(*Platycladus orientalis* (L.) Franco)[4,5]。也有人将电容法应用于林木树势的测定[6]。然而,应用热红外成像技术直接检测林木活力或健康状态的报道至今尚不多见。

应用热红外成像技术,我们曾研究过一些树木枝叶的温度变化[7]、水分状态[8]以及树叶的冻害症状等[9]。本研究用热红外成像技术检测了侧柏伐根面以及生长锥木芯之边材与心材的水分和能量代谢关系。从某种意义上讲,红外热温值更多地反映了树液的状态而不是木材的比例。决定树木(尤其是侧柏)健康状态的是边材和心材的树液比或红外热温比。该检测手段有潜力成为检测侧柏林木活力或健康状态的重要指标。

侧柏为我国华北地区荒山(尤其是石灰岩山地)造林的先锋树种,也是城镇园林绿化的重要树种资源。由于侧柏林大多是以防护林为目标林种的生态公益林,立地条件差、密度过大、抚育管理较为粗放;少数林分,尤其是一些都市和名胜地的景观林,由于林龄偏大、树势衰退,导致一些弱寄生性虫害时有发生[10,11,12],柏肤小蠹(*Phloeosinus aubei*,Perris)和双条杉天牛(*Semanotus bifasciatus*,Motschulsky)等蛀干害虫是常见的侧柏衰弱

* 资助项目:国家自然科学基金(31170671);山东省科学技术发展计划项目:2012GNC11107;通讯作者:王斐,1959年生,山东省林业科学研究院研究员。

木、濒死木、枯立木、伐倒木和新栽幼树上寄生的害虫[13,14,15]。应用热红外成像技术，若能成功地检测侧柏树木的活力状态和受害状况，将对侧柏生态公益林的科学管理和健康成长发挥重要作用。

2. 材料与方法

2.1 研究材料

本研究以山东省淄博市沂源县亳山林场唐山景区和山东省林业科学研究院燕子山试验林场的侧柏生态公益林为研究对象。其中，唐山侧柏林分位于沂源县唐山寺周边的侧柏纯林，林龄 80 多年，林分相对稀疏。燕子山林场的侧柏林分始建于 20 世纪 50 年代，目前每公顷有侧柏 3000 株以上。

2.2 热红外图像拍摄与分析

红外热像用 NEC H2640 型热红外相机（波长 8～13 μm）拍摄。该相机的测温范围为 −40～500 ℃，最小感温能力 0.03 ℃。红外热像是在设定灵敏度自动追踪、发射系数为 0.98 的状态下手持摄像机于目标叶/木芯上方 50～100 cm 处调焦清晰后，在自然光下顺光拍摄的。拍摄鳞叶时将 30～40 叶隆成束，直接拍摄成束的鳞叶部位。树干的木芯是用 5 mm 生长锥在胸高部位钻取的，木芯以钻过树干的一半为准。木芯钻取后立即放入适当的塑料吸管中封闭两端待用，采样完成后所有木芯迅速整齐地用双面胶带粘在画板上，随后持续拍摄热红外图像，直至拍到清晰的图像为止。

林地的热像温度拍摄于上午 9:00～11:00，拍摄时手持摄像机于距地面 160 cm 处调焦清晰后，在自然光下顺光拍摄，热像温度值来自整幅热像的温度平均值。

2.3 相对热温指数

为了增加不同热像中热温数据的可比性，在热温数据分析时有时应用相对热温指数。相对热温指数是目标器官或组织之间的热温差或热温比值，如指温差指数和指温比指数，其定义分别如公式(1)和公式(2)所示：

$$TD_{lf} = \sum_{i=1}^{n}(Tf_i - Tl_i) \quad (1)$$

式中：TD_{lf} 为温差指数，Tl_i 为第 i 次重复观测的热温数值，Tf_i 为第 i 次重复观测的手指热温数值，n 为重复数。

$$TR_{lf} = \sum_{i=1}^{n}(Tl_i / Tf_i) \quad (2)$$

式中：TR_{lf} 为温比指数，Tl_i 为第 i 次重复观测的热温数值，Tf_i 为第 i 次重复观测的手指热温数值，n 为重复数。

此类热温指数以观测者的指温为参照，测试结果更加稳定和具有可比性。在实际观测过程中，有时也应用同一热像中不同部位的热温比值进行比较分析，如伐倒木之伐桩横截面的边材和心材热温比(边心温比)。

2.4 蛀干害虫的鉴别和数据分析

柏肤小蠹和双条杉天牛的鉴别方法是采集具有羽化孔的样木段或树皮，以在光学显微镜(Caikon DMM 300C 型)低倍放大率下观察到相应的成虫和幼虫为准。

研究结果的数据统计、分析及制图是在 Microsoft Excel 2003 和 Kaleida Graph 4.0 软件中进行和完成的。

3. 结果与分析

2013 年冬和 2014 年春，在淄博市沂源县亳山林场唐山景区的 80 多年生侧柏林分中发现一些植株明显是衰弱木，尤其是在土壤瘠薄、立地条件较差的林分中。在观光游览区外来虫源概率大的地方发生了局部的柏肤小蠹和双条杉天牛的侵害，受害植株的树干上可以看到清晰的成虫羽化孔及树干流脂现象。为确定受害木的活力，我们选取有代表性的成龄和幼龄植株，在树干胸高处用 5 mm 生长锥钻取样条，并将这些样条一字排列成行(见图 1b)，应用热红外成像仪观测其边材热像温度(见图 1a)。结果表明：柏肤小蠹和双条杉天牛受害木(见图 1a 左侧 1～10)的边材热温明显高于正常生长的侧柏林木(见图 1a 右侧 11～20)，并存在统计学意义上的极显著差异(图 1c，$F=13.2^{**}$，$p=0.0016<0.01$)。从受害木的热像可以看出，一些枯立木的边材热温与心材难以区分(见图 1a 左侧 5)，一些衰弱木的边材热温与心材相差无几(见图 1a 左侧 10)，而活力旺盛的林木边材热温普遍较低(见图 1a 右侧 11～20)。这说明，与正常侧柏林木相比，此类蛀干害虫受害木树液输导系统的功能相对较弱。

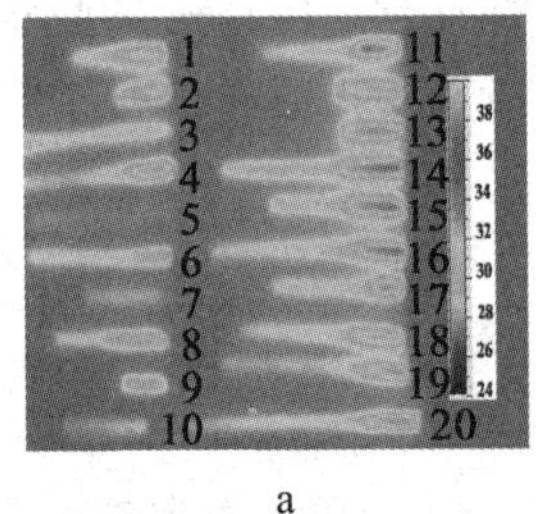

a

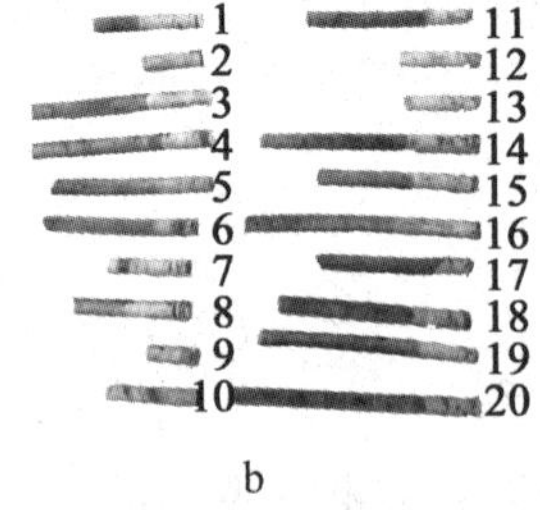

b

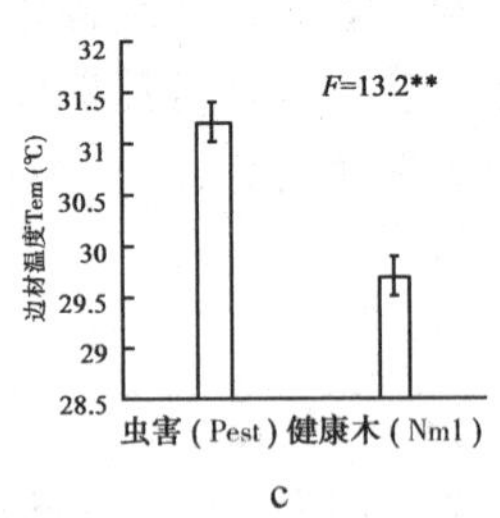

c

图 1 柏肤小蠹和双条杉天牛侵袭的侧柏主干生长锥芯之热像、数字图像和边材热像温度的对比：a. 受害木(左 1～10)和正常对照木(右 11～20)的热红外图像；b. 图像 a 中所对应的数字图像；c. 受害木(Pest)和正常木(Nml)边材温度(Tem)对比柱状图(参见彩色附图 6)。

我们在山东省林业科学研究院燕子山试验林场进行的林窗补植试验研究结果表明：由于立地条件等环境因素的巨大差异，新栽苗木分化严重（见图 2a）。一些植株从移植开始一直处于受胁迫状态，鳞叶颜色呈黄绿或黄棕色，热像温度明显偏高（见图 2a-受胁迫），指温差值为负值，指温比最高且大于 1。尽管另一些新植苗木叶色为正常绿色（见图 2a-正常），与天然更新的苗木（见图 2a-自然更新）相比，鳞叶的热像温度要高；而自然更新的幼树和幼苗的根系未受移植损伤，水分代谢平衡，叶色呈暗绿色，热像温度最低。其中处于胁迫状态的侧柏新植苗在环境优越的条件下尚可恢复其活性而逐渐降温复绿，但在较差的环境条件下，常因受过度胁迫而枯萎。正常的移栽苗木经过一定的缓苗期大多能够成活下来，尤其是在雨量充沛的雨季栽植的植株。但是，新栽苗木和自然更新苗之间的热温差异甚至在栽培的第二年仍能观测出来。这可以解释为什么柏肤小蠹和双条杉天牛易侵害新栽的幼树这一现象，也说明天然更新是恢复侧柏林生机的可靠途径。

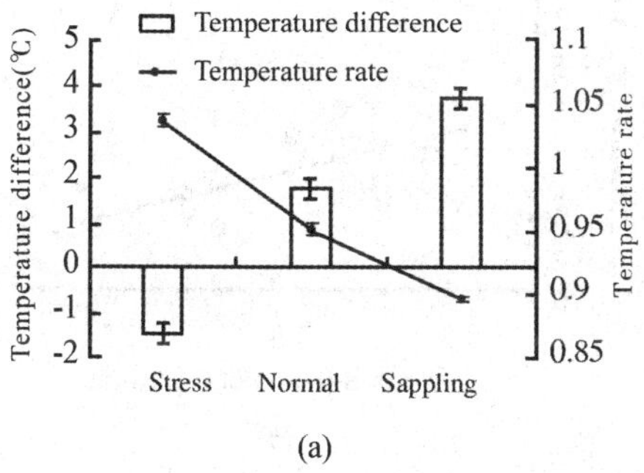

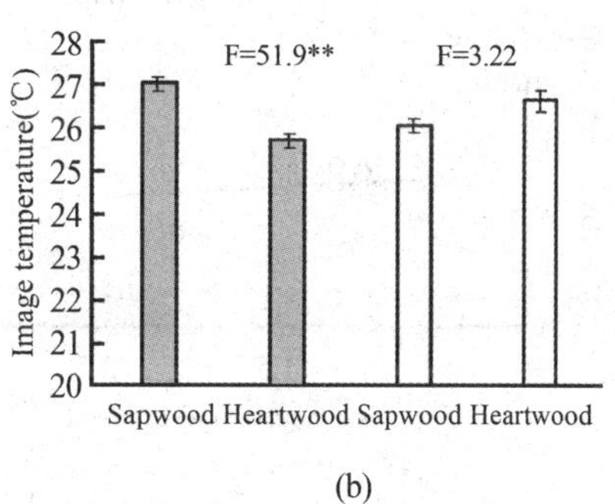

图 2 侧柏林窗新栽苗木的鳞叶指温差（Temperature difference）、指温比（Temperature rate）以及侧柏树干横截面边材与心材热象温度的对比。图(a)为侧柏林窗新栽苗木指温差、指温比，其中包括植后一直枯黄的苗木（Stress）、绿色的苗木（Normal）和自然更新的幼苗（Sapling）；图(b)为过密林分中严重分化之侧柏幼树中的正常植株（深色柱）和衰弱木（无色柱）之边材（Sapwood）与心材（Heartwood）热温比较。

在燕子山过密（超过 3000 株/公顷）的侧柏林分中，侧柏幼树常因受压而衰退，以至于个体之间分化严重。一些植株成功透过树冠间隙而长到树冠层；另一些则严重受压而弯曲、衰退甚至枯萎。用热红外成像仪进行观测对比表明，正常侧柏林木的边材与心材之间热像温度差异显著（见图 2b-深色柱），而衰弱个体的边材与心材之间往往没有显著的差异（见图 2b-无色柱）。

侧柏在幼龄阶段喜光、喜湿润环境的习性使得更新苗木集中在母树的下方，结果导致林分密度过大，个体生存空间狭窄；再加上侧柏下部枝条容易枯萎的特点，使树冠变小，树势衰退。直接表现为植株的树冠大小与边心温比之间存在显著的线性反相关关系（见图 3a，$p<0.01$），这显然与 20 年前韦林提出的“活力指数”理论相吻合。也就是说，林分过密、种内竞争激烈会导致侧柏个体活力低下，林分健康状态偏低。

另一方面，边心材面积比与树冠冠幅、冠长以及边心温比之间没有明显的相关性（见图 3b、图 3c、图 3d）。这表明，边材-心材面积比更多地反映了树干的木质结构，而边心温比反映的是树干截面水分和能量分布的特征，从而揭示了树干截面上有效的树液输导面积。

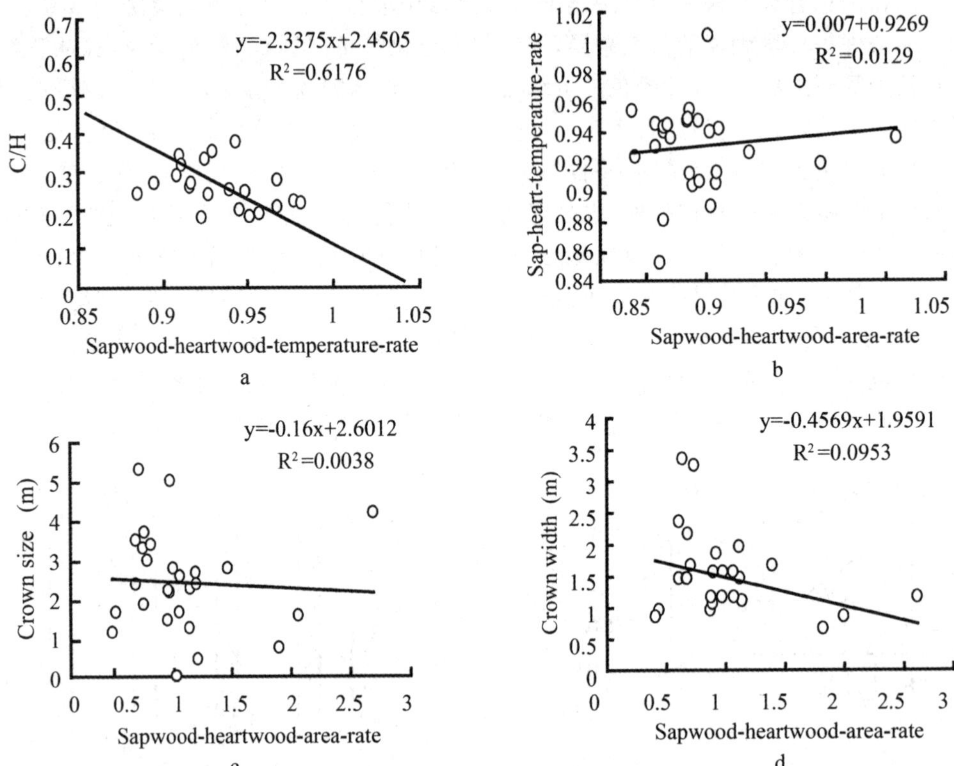

图 3　相关分析结果：图 a 为侧柏树干边心热温比与冠幅一冠长比(C/H)之间的相关关系；图 b 为侧柏树干边心面积比(sapwood-heartwood-area-rate)与边心热温比(sapwood-heartwood-temperature-rate)间的相关关系；图 c 为侧柏树干边心面积比与冠长(Crown size)间的相关关系；图 d 为侧柏树干边材面积比与冠幅(crown width)间的相关关系。

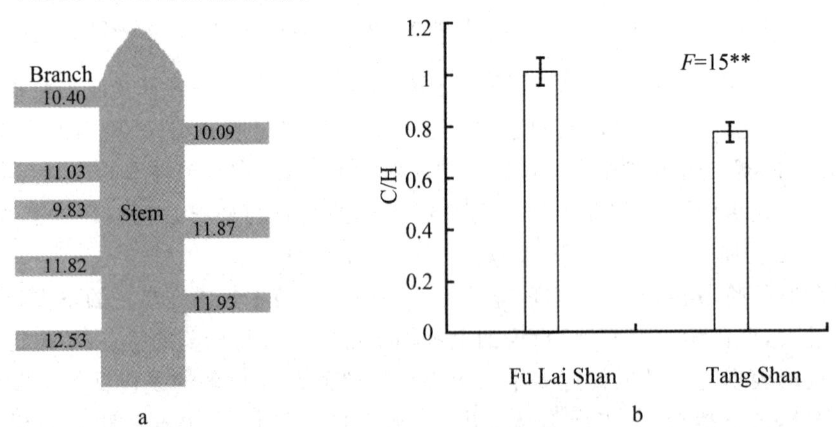

图 4　反映侧柏健康状态的两个指标：图 a 为枝条(Branch)热温在树干(Stem)上的空间分布示意图；图 b 为浮来山(Fu Lai Shan)和唐山(Tang Shan)之侧柏的冠幅/冠中高之比。

侧柏枝叶自下而上枯萎的习性是其适应极端环境的一种重要手段，以至于在过密的林分中 7～8 m 高的侧柏植株仅靠几十甚至十几厘米长的树冠来维持生存。输导系统的阻塞用热红外检测主干下部侧枝切口时会表现出明显的高热温。相比之下，若上部枝条保持较强的生命力，则树液输导活跃可维持较低的热温（见图 4a）。侧柏在遭受极端干旱等环境胁迫之后，如果将鳞叶自下而上分成三组，与图 2a 所示的幼苗幼树相似，则热温差值往往由低到高排列，而热温比则由高到低变化。

在侧柏集中栽培的石灰岩山地地区，因林分过密、立地条件差、风口逆境胁迫等因素，常导致树冠呈倒三角形（正常情况下健康植株的树冠为塔形）。与山东省东南地区的浮来山相比，沂源唐山景区的侧柏林龄偏大，立地条件偏差，导致二者树冠冠形差别明显：唐山侧柏林的冠幅与冠中高（树干基部到树冠最宽部位的高度）之比明显低于浮来山的侧柏林木（见图 4b），这似乎是该地发生柏肤小蠹和双条杉天牛等蛀干害虫的重要条件。因此，采取必要的营林措施，防止侧柏生态公益林分亚健康状态的出现是至关重要的。

侧柏林分密度较大，常形成单层纯林。林内环境条件不利于天然更新，林木常随树龄增长而衰退。在山地阳坡的侧柏林窗内，侧柏更新苗木的数量与局部环境有关。在侧方庇荫的阴湿环境条件下测得的林地热温较低，侧柏天然更新容易，更新苗数量多（见图 5A），生长快而健壮。在空旷裸露的干热环境中测得的热温最高，侧柏天然更新困难，更新苗少见（见图 5B）；而在侧方光照充足、相对干旱的上方庇荫环境中测得的热温偏高，尽管有侧柏天然更新苗生长，但数量较少且生长偏弱（见图 5C）。显然，侧柏幼苗更新需要最适宜的环境条件。合理的密度控制是侧柏林更新、个体复壮、恢复生机的重要手段。欲恢复植株的活力和森林的健康，关键在于适宜的天然更新机制，而热红外成像技术在检测树木的健康状态和幼苗适生环境中应该能发挥重要的作用。

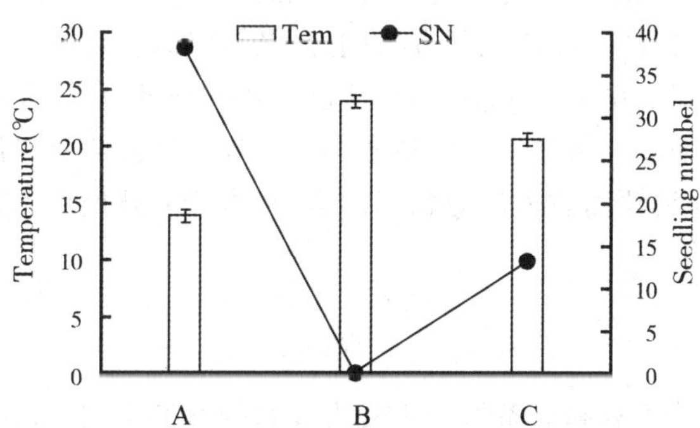

图 5　典型的阳坡林窗内不同区域侧柏更新苗数量（SN）以及从苗木上方向下测得的热像温度值（Tem）。其中，A 为侧方庇荫区域，B 为裸露空旷区域，C 为侧方光照充足的区域。

4. 讨论

本研究表明，热红外成像技术的应用具有早期检测侧柏衰弱木和虫害木的潜力，为防止柏肤小蠹和双条杉天牛等蛀干害虫的大发生提供必要的监测信息。在燕子山试验林场的过密侧柏林分中，林木密度超过 3000 株/公顷，林木分化严重。即使存在弯曲木、衰退木和枯立木，也未发现侧柏柏肤小蠹和双条杉天牛的蔓延。这说明柏肤小蠹和双条杉天牛等蛀干害虫传播能力低，由于没有虫源而未发生和流行，这与此类蛀干害虫常发生在人为活动频繁的地区而没有人烟的偏远林区几乎没有虫害踪迹的事实相吻合[14,15]。也间接说明，侧柏柏肤小蠹和双条杉天牛等蛀干害虫发生的条件一是有侵染的虫源，二是有大量衰弱木，两者同时存在虫害才能发生。结合营林抚育措施恢复侧柏林分健康，同时控制虫源的传播是综合防控此类虫害大发生的重要途径和手段。

另一方面，边材的水分代谢和流脂能力等物理因素或许也是侧柏柏肤小蠹和双条杉天牛等昆虫选择性的制约因素。这些害虫只发生在衰弱木、濒死木和枯立木上，这与此类害虫成虫的选择性产卵有关。这种选择性除了与树木的化学分泌物有关外，树体内边材树液少，发射热红外光谱量大等物理因素或许起着不可忽视的作用，因此有潜力应用于侧柏柏肤小蠹和双条杉天牛等蛀干害虫的评测并成为此类害虫综合防治的有力工具。

侧柏苗木在母树下更新的特点使得侧柏林分常形成过密纯林。伴随着这些林分的成长严酷的种内竞争造成林木严重分化，以至于使一些树木衰退甚至枯萎。要从根本上解决此问题，必须加强对侧柏生态公益林的抚育管理，关键在于促进侧柏的天然更新并恢复林分的生机，减少亚健康木和衰退木的数量，提高其健康水平。热红外成像技术对衰弱木和林地水分状态的敏感性特点使其有潜力应用于新栽苗木活力状态的检测、林分健康状态的诊断以及侧柏更新林地环境的评价等。为此，在热红外成像设备监测的基础上，合理的密度控制、立地条件选择和管理以及在干旱瘠薄山地阳坡环境下适度的水分供给和调节有利于侧柏生态公益林的更新和维持。

尽管应用树木边材面积比等指标对森林健康评测起着不可忽视的作用，但不同健康状态的树木之间能够有效地输导树液的边材面积的不同往往会使这种评价产生偏差。此外，单纯探测边材部位树液流速的观测方法同样不够准确，因为树木的生长状态与边材中可有效输导树液的组织关系更加密切，而且伴随着侧柏树木活力的衰减其有效输导树液的边材面积在不断地缩减。热红外成像技术能够很好地探测边材树液输导的有效部位和能力，因此有更加可靠地评测树木活力状态的潜力，本研究就提供了这样的一个例证。

5. 致谢

本研究中侧柏柏肤小蠹和双条杉天牛受害症状的鉴别得到了山东省林业科学研究院王西南研究员和武海卫博士的帮助，在此表示衷心的感谢。

主要参考文献

[1]Waring R. H., Thies W. G., Muscato D. Forest Science, 1980, 26: 112-117.

[2] Gao Jun-Kai. Docter Thesis, 2007, Beijing Forestry University.

[3] Wang Yan-Hui, Xiao Wen-Fa, Zhang Xing-Yao. 13-years Succession Dynamic of Kunyushan Natural Forest: Change of Diversity, Species Turnover and Herbivorous Insect's Short-term Disturbance[J]. Scientia Silvae Sinicae, 2007, 43(7): 78-85.

[4] WangHua-Tian, Zhao Wen-Fei, Ma Fu-Yi. Scientia Silvae Sinicae, 2006, 42(7): 21-27.

[5] Liu De-Liang, Li Ji-Yao, Ma Da. Chinese Journal of Ecology, 2008, 27(8): 1262-1268.

[6] Li Xing-Wei, Zhou Zhang-Yi, Zhang Jun-Lou. Guangdong Forest Science andTechnology, 2002, 18(1): 19-24.

[7]Wang F., Yamamoto H., Ibaraki Y., Iyayak, Yoshikoshi H., Takayama N. Journal of Agricultural Meteorology, 2010, 66(2): 81-90.

[8] Wang F. Journal of Forestry Research, 2010,21(4): 465-468.

[9] Wang F.,Omasa K., Xing S. J., Dong Y. F. Ecological Informatics, 2013, 16 (2013): 35-40.

[10] Yu Su-Ying, Li Yu-Ying. Hebei Forest Science andTechnology. 2000, 5: 26-28.

[11]Jia Sui-Tai, Xi Zhong-Cheng. Chinese Forestry, 2008, 6: 44.

[12] Liu Yu-Jian. Master Thesis,Beijing Forestry University, 2005, 6.

[13] Wu Xiao-Ying. Doctor Thesis, Beijing Forestry University,2009.

[14] Sun Yue-Qin. Master Thesis. Beijing Forestry University, 2000.

[15]Niu Guang-Pu. Forest Pest and Disease, 2008, 27(4): 15-17.

(原文发表于《光谱学与光谱分析》2015,35(12):3410-3415.)

Thermographic analysis of leaf water and energy information of Japanese spindle and glossy privet trees in low temperature environment*

WANG Fei[1], Kenji Omasa[2], XING Shanjun[1], DONG Yufeng[1]

(1. Shandong Forestry Research Academy, 250014, Jinan, China; 2. Graduate School of Agriculture and Life Sciences, the University of Tokyo, 113-8657, Japan)

1 Introduction

The specific heat and the latent heat of the melting and evaporating process of water lays the foundation of transpiration cooling (Clements, 1934; Rosenberg, 1974). With the highest specific heat, water tends to stabilize temperature, and this process is reflected in the relatively uniform temperature of islands and lands near a large body of water (Rosenberg, 1974; Kramer, 1983). Transpiration causes the temperature of trees to vary less than the air temperature (Gates, 1968). The lower temperature of deeper soil water and soil water under the coverage of crown self increases the effectiveness of the transpiration cooling of trees (Rosenberg, 1974). The soil temperature amplitude decreases with increasing depth during both summer and winter. At a depth of 40 cm, the temperature wave is significantly damped, particularly in winter, whereas no significant diurnal/annual wave is found at a depth of 80 cm (Rosenberg, 1974). In Jinan, the maximum difference between the air temperature and the deeper soil temperature reaches more than 10 ℃ in the summer and more than 5 ℃ in the winter. Therefore, the water under a deep layer of soil results in the slight variation observed in the temperature of trees compared with the air temperature.

However, plant transpiration has been considered an unavoidable evil for a long pe-

* Corresponding Author: WANG Fei, Email: wf-126@126.com, Tel: 0086-0531-88557594, Fax: 0086-0531-88932824. This work was supported by the National Natural Science Funds of China (ID Number: 31170671) and the Project of Science and Technology Development in Shandong, China (ID Number: 2012GGB2201)

riod of time (Kramer, 1983). The active action of water in the transpiration cooling of plants has even been disregarded to some extent, and the warming action of sap water in winter has not been analyzed. During the winter, the warm sap from deep soil plays an important role in the cold hardness of many plants, especially in Jinan, China. In this paper, we evidenced the "sap warming" process of some evergreen tree/shrub species during a cold Jinan winter using thermography. The proper consistency between the leaf tip and margin scorched areas and the lower temperatures in the same area suggest that this type of symptom results from a lack of warm water from the root system.

Various noises and small temperature differences in field measurements make it difficult to compare different thermal images (Chaerle *et al.*, 1999; Chaerle and Van-Der-Straeten, 2000; Grant *et al.*, 2006). To increase the comparability of this type of images, some researchers have attempted to use contrast models (Jones and Leinonen, 2003) in thermographic detection. Nilsson (1995) observed a significant leaf temperature decrease in a gust of wind, which implies that the dynamics of the imaging temperature is important for the identification of the stress status of plants. In this study, hand-heated leaves were used to study the leaf temperature and the sap warming phenomenon in a winter environment. The significant temperature difference on the heated leaf made it easy to obtain the thermal image and reduced the systematical error. In this study, the temperatures of the leaves and twigs of Japanese spindle and glossy privet were detected using thermography during the cooling process after hand heating. The significantly amplified variation in the imaging temperature between the normal and severed parts of leaves and branches during the cooling process after hand heating indicated that it was possible to detect the freezing dehydration of leaves, the scorching of leaves and the branch dieback by analyzing their changes using thermography. Moreover, these freezing stresses can be early detected before the appearance of any visible symptoms.

2 Materials and methods

2.1 Background of the study

This study was performed from December 2012 to January 2013 near the Yanshan crossroads in Jinan, China, which is located at E117° 0′ 0″ and N36° 24′ 0″. This region has an extreme minimum temperature in January of −14.5 ℃. During the study, the maximum and minimum temperatures were 8 ℃ and −13 ℃, respectively. Three rounds of snowfall occurred on the days of Dec. 13−14, 20−21 and 28 in 2012, which resulted in 12.0, 9.6 and 7.0 mm, respectively, of precipitation. In December of 2012, the total precipitation was 29.2 mm, which is the second highest amount of monthly precipitation in the years from 1951 to 2012. Therefore, it can be stated that the studied winter was a durative freezing winter. During this period, several green hedge stocks and roadside trees of Japanese spindle (*Euonymus japonicus*, Thunb.) and glossy privet (*Ligustrum lucidum*, Ait.) were studied. Some Japanese spindle specimens under the bridge of the Yanshan crossroads, a place which received almost no water of rainfall or snowfall, were studied, compared with the normal specimens in the field.

2.2 Imaging temperature measured by thermography

The leaf, stem and branch kerf temperatures were determined using thermography technology (Jones, 1999; Jones *et al.*, 2002; Jones and Leinonen, 2003; Prytz *et al.*, 2003) with a NEC H2640 thermal infrared (8—13 μm) camera with a temperature measuring scope that ranged from −40—500 ℃ and a minimum sensible temperature of 0.03 ℃. Throughout the measurements, the camera was handheld approximately 50 cm above the objective leaves/branches and focused to clarify the image. Smoothly expanded leaves were selected to avoid the systematic errors that are obtained in the collection of thermal images. Thermal images were obtained during the processes of hand-heating and non-hand-heating in the LVT automatic sensitivity tracing mode. After the target leaf was clamped between two hands and heated for approximately 10 seconds, thermal images were continuously taken as the temperature decreased until a temperature that was nearly equal to the around environmental temperature. The clearest thermal image was used to analyze the image temperature. The imaging temperatures were determined using the InfReC Analyzer NS9500 software provided with the camera. The difference in the temperature (Dt) obtained from the thermal images of each duplicate was the difference between the average temperature and the minimum temperature of the duplicate. This value was calculated using Equation (1):

$$Dt_i = (T_i - T_{min}) \tag{1}$$

where Dt_i is the difference in the temperature of the ith duplicate ($i=1, 2, \ldots, m$, where m is the number of duplicates), T_{min} is the minimum temperature value of the ith duplicate, and T_i is the average temperature value of the ith duplicate.

2.3 Analysis of RGB images

In the study, the leaves and stems were monitored during the process of winter freezing from December of 2012 to January of 2013 through making both thermal and RGB images. The RGB images were obtained with a CCD camera (Fuji SL 305). These images were stored as tiff files. No special constraint was used in the determination of the leaf scorch area percent (LSAP), with the exception of blurry images. The LSAP values were defined and calculated using the method described by Wang and Omasa (2012).

2.4 Definition and calculation of the leaf angle between the petiole and the tip

The leaf angle between the petiole and the tip (LAPT) of the target leaves is a side-glance bending angle between the leaf petiole and the leaf tip (Fig. 1), which was measured using the RGB images obtained with the CCD camera (Fuji SL 305). This value was measured using the angle analysis tool in the UTHSCSA Image Tool 3.00 software. The LAPT was directly measured if it was a sharp angle (Fig. 1a), indirectly determined according to its corresponding angle if it was an obtuse angle (Fig. 1b) and directly measured according to the point of the leaf tip if it was a reflex angle (Fig. 1c).

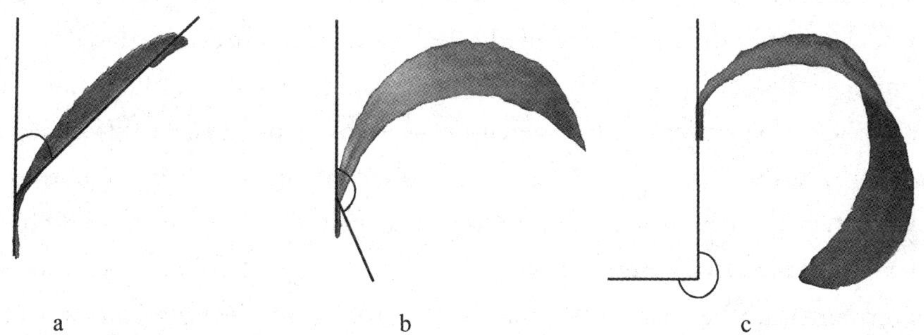

Fig. 1 Leaf angle between the petiole and the tip (LAPT):
a. sharp angle; b. obtuse angle; c. reflex angle.

2.5 Leaf water content

The water content (*WC*) of glossy privet leaves was differentially measured to compare the *WC*s in the leaf tip and the base. During the study, scorched leaves were separated into two sections: tip and base. The water contents of these sections were measured using the rapid weighing method with a 1/10000 g Shimadzu electronic balance (AUY 120) under room temperature conditions. The weight of the sampled leaf sections was weighed immediately after the sample was obtained from the field. After the dry weight was obtained, the water content was calculated using Equation (2):

$$WC\% = \frac{FW - DW}{FW} \times 100 \tag{2}$$

where *WC* is the water content, *FW* is the fresh weight of the samples and *DW* is the dry weight of the samples.

3 Results and analysis

3.1 Image temperatures in a normal winter environment

Every object continually exchanges energy with its environment and tends to maintain an energy balance. Plants attempt to always maintain their temperatures to equal to that of the surrounding environment; this is particularly true for the surface temperature after a persistent energy exchange (Fig. 2a). Therefore, the surface temperature of leaves (or different parts of the same leaf) usually does not significantly change in winter, especially at night and during cloudy days. All the measurements do not exhibit statistically significant difference between the leaf tip and the base area (Fig. 2b). However, a higher temperature was occasionally found at the leaf base in the extreme cold environment, although the repeatability of this result was not high. In this study, the leaves of Japanese spindle and glossy privet were used as typical examples (Figs. 3a and 3b). The point temperature variance of the leaves usually ranged within 1.0 ℃. However, during a sudden change in the environment, a significantly higher temperature was measured at the leaf base or the leaf venation system. This type of difference appeared to be the result of the high inner temperature of plant bodies. This hypothesis can be proven by the fact that the branch kerf of Japanese spindle (Fig. 3c-1) exhibited a higher instant temperature than the stem surface (Fig. 3c-2). This difference between the inner temperature and the surface temperature is often statistically significant (Fig. 2c), especially in an environment with an extremely low temperature and snow. The large temperature variance between the inner part of plants and the environment results in its easy detection by thermography. The sap that brings relatively warm water from underground to the terminal leaves maintains the persistent difference between the inner and the surface temperature. This finding is evidenced by the persistently higher kerf temperature in the lower parts of the stem (Fig. 2a -●-●) compared with the upper/terminal part (Fig. 2a -◇-◇) in Japanese spindle.

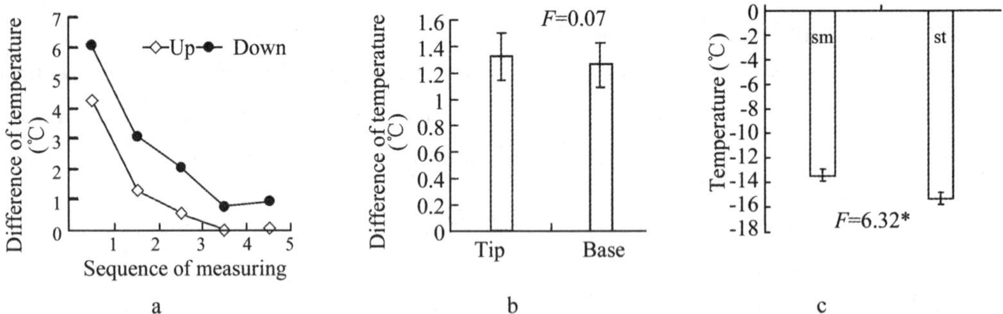

Fig. 2 Statistical results of leaf and branch kerf temperatures. a. Temporal series of temperature differences (mean-minimum) between upper branch kerfs (Up) and lower branch kerfs (Down); these temperature measurements show a lag characteristic of the decreasing temperature in the lower branches. b. Temperature difference (mean-minimum) between the tip and the base area of glossy privet leaves; the differences in the temperatures of the tips and bases are not statistically significant. c. Significant temperature difference between branch kerfs (sm) and main stems (st) of Japanese spindle seedlings.

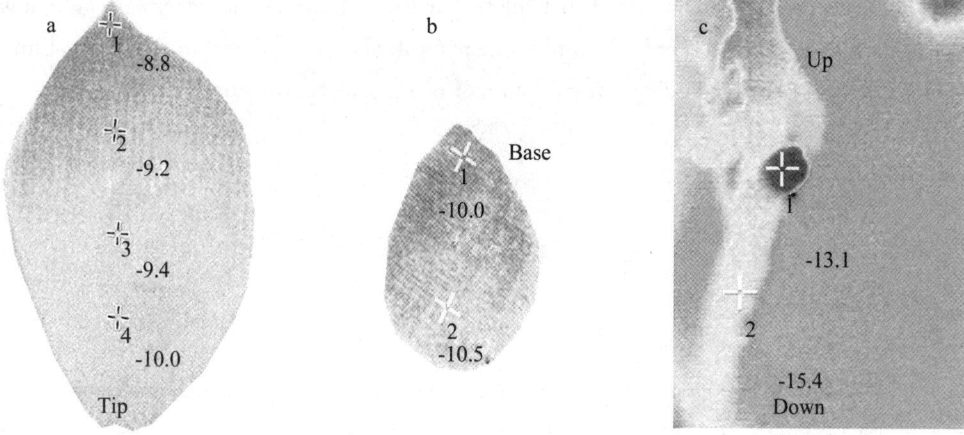

Fig. 3 Winter thermal images of the main stem and leaves of Japanese spindle and glossy privet. a. Thermal image of glossy privet leaf with gradient (high to low) temperatures from the base to the tip (a-1, a-2, a-3 and a-4). b. Thermal image of Japanese spindle with temperature (high to low) gradient from the base to the tip (b-1, b-2). The centers of the crosses indicate the point at which the temperature was measured. c. Main stem of a Japanese spindle seedling with a high-temperature branch kerf (c-1) and a low-temperature stem (c-2).

3.2 Thermatic analysis of leaf water and thermal status after hand heating

In normal winter environments, the temperature difference within a leaf blade is not easily measured (Fig. 4a), although a warmer leaf vein system is sometimes detected using thermography. The noise from the reflecting light of the leaf surface often resulted in a significantly large measurement error, particularly in the measurement of leaves with a bright luster, such as the surface of the Japanese spindle plant. To avoid this type of noise, we used the sunshine heating method to obtain thermal images with a significant temperature difference between different plant parts (Wang *et al*, 2010) in the summer. In the present study, the hand-heating method was used in the thermal detection process. The temperature of the target leaves, which were heated by the observer's hand and thus subjected to a constant energy source of 36.8 ℃, can reach approximately 20 ℃. This heating resulted in clearer thermal images and a significant temperature difference between the different parts of leaves exposed to varied stress conditions. After hand heating, the temperature difference within a leaf can reach 3—4 ℃ (Fig. 4b), and 5—8 ℃ or more under extreme conditions, particularly in leaves with scorched tip and/or margin (Fig. 4c). Most of the thermal images obtained during the cooling process after hand heating showed a clear major vein system with a higher temperature (Figs. 4b and 4c) through the leaf, with the exception of a partial area with abruptly thinner veins. The thin leaf venation at the tip and margin of Japanese spindle leaves is a special example (Figs. 5a, 5b and 5c). In this special case, the higher temperature area is consistent with the area that presents a significant trace of water (Figs.

5d and 5e). This result demonstrated that the higher temperature of the major vein system is the result of its higher water content. This phenomenon can also be observed in the thermal imaging of a crosscut Japanese spindle leaf (more than half of the leaf is crosscut; Fig. 5f).

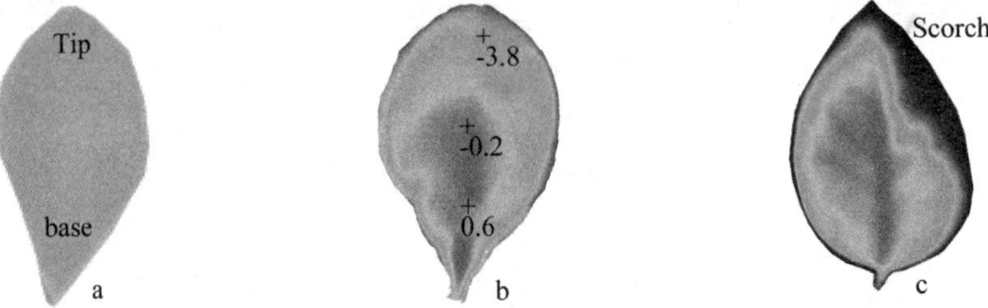

Fig. 4 Thermal images of Japanese spindle and glossy privet leaves as the temperature decreased after hand heating. a. Thermal image of a normal Japanese spindle leaf. b. Thermal image of a Japanese spindle leaf before the appearance of leaf scorch symptoms. c. Thermal image of a glossy privet leaf after the appearance of leaf scorch symptoms.

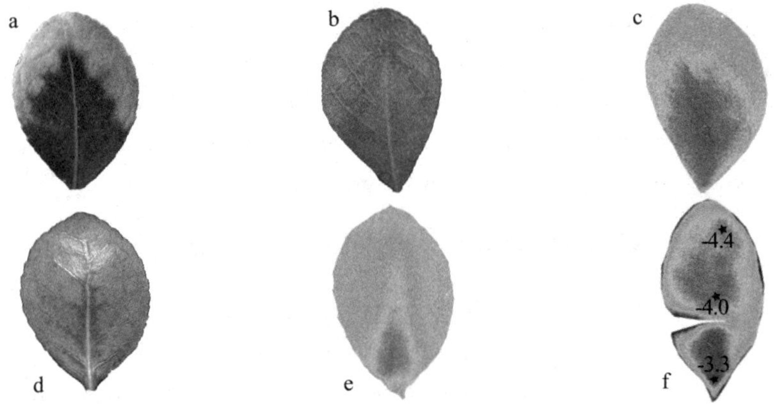

Fig. 5 RGB and thermal images of the front and back surfaces of a Japanese spindle leaf as the temperature decreased after hand heating. a. Front surface of a leaf with a clear tip and a margin scorch. b. Back surface of the leaf in (a) shows the thinner major or minor vein in the tip area. c. Thermal image of the leaf in (a) with almost the same low temperature at the tip and margin areas. d. A leaf with significant water trance at the base of the leaf and along the major vein. e. Thermal image of the leaf in (d) as the temperature decreased after hand heating. f. Thermal image of a Japanese spindle leaf, more than half of which was crosscut(See attached color Fig. 7).

3.3 Identifying the leaf tip and margin scorches using thermal images

A statistically significant temperature difference between the scorched tip and the living base area of scorched leaves of Japanese spindle and glossy privet was measured after hand heating (Fig. 6a; $F=5.43$ and $P<0.05$). The scorched area of glossy privet measured using both thermal and RGB images of the same leaves showed no significant difference (Fig. 6b; $F=0.0002$ and $P>0.1$). However, the typical temperature difference and temperature changes can be found between the living leaves and the wilted leaves of Japanese spindle after hand heating. Due to their low water content, the wilted leaves usually exhibit a higher temperature at the beginning of the measurement and a lower temperature at the end of the measurement compared with living leaves (Fig. 6c). After the scorched area is separated from the living area of glossy privet leaves, the scorched tip and margin areas exhibited a similar lower temperature and water content (Figs. 7c and 7d), whereas the living base area presented a higher temperature and a higher water content (Figs. 7c and 7d). Therefore, the hand-heating method can be potentially used to distinguish between dead and living stocks and between normal and scorched leaves of Japanese spindle and glossy privet.

The appearance of scorched symptoms in the tips and margins of the leaves of many tree species is related to their water metabolism (Wang *et al.*, 2009a, 2009b; Wang and Omasa, 2012), particularly during periods of summer drought. Similarly, the scorched patterns observed in the tips and margins of Japanese spindle during the winter freezing dehydration process appear to be also related to a water imbalance. In this study, an evident leaf bend in the Japanese spindle was determined through the measurement of the LAPT (Fig. 7a; $F=147.8$ and $P<0.01$) during the freezing process. As the air temperature was rose to above 0 ℃, the bended Japanese spindle leaves usually return to their normal turgid status (Fig. 7a) with a small LAPT. We also found a similar leaf bending phenomenon in detached Japanese spindle leaves in spring and measured a similar LAPT value (Fig. 7b). Moreover, more severe leaf tip and margin scorches were found in the leaves of Japanese spindle trees planted in dry and cement-polluted soil conditions compared with those found in the normal conditions observed in Jinan, China. Newly transplanted seedlings also exhibited the described scorch symptoms due to their imperfect root systems. Therefore, these findings suggest that the leaf tip and margin scorches found in Japanese spindle in the freezing conditions of winter are related to a water imbalance and/or a lack of sap warming.

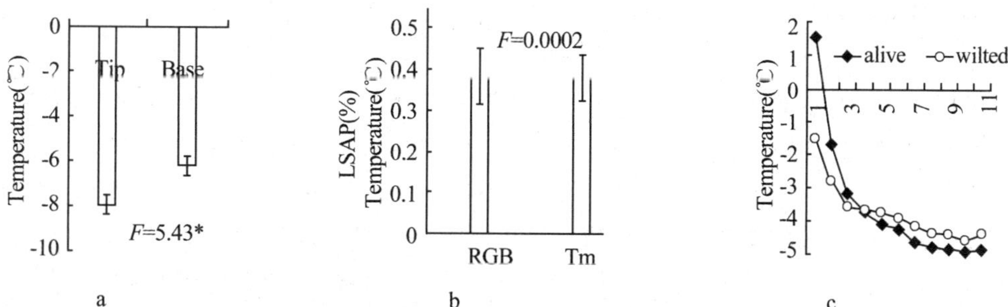

Fig. 6 Analysis of the thermal and RGB imaging of Japanese spindle. a. Temperatures of the tip and base leaf sections as the temperature decreased after hand heating. b. Comparison between the leaf scorch area percent (LSAP) obtained from the RGB images (RGB) with that obtained from the thermal images (Tm). c. Comparison of the temperature of living and wilted leaves of Japanese spindle.

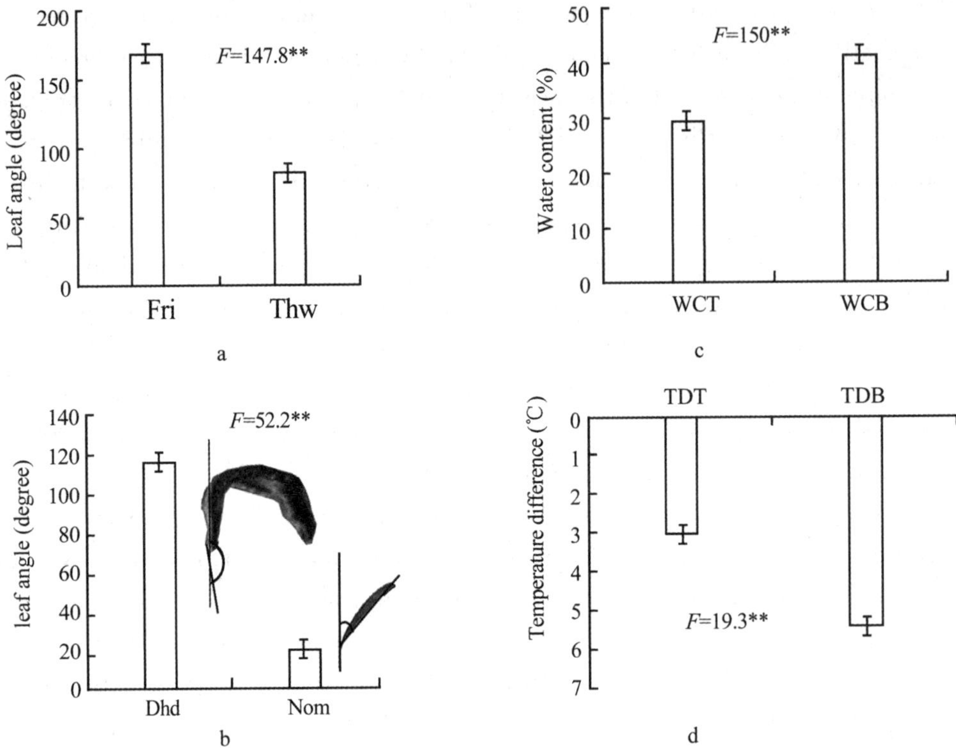

Fig. 7 Leaf angle of Japanese spindle and water content and temperature differences in glossy privet leaves. a. Leaf angle between the petiole and the tip of freezing-dehydrated Japanese spindle leaves (Fri) and thawing turgid leaves (Thw) in winter. b. Leaf angle between the petiole and the tip of air-dehydrated Japanese spindle leaves (Dhd) and normal turgid leaves (Nom) in spring. c. Water content of glossy privet leaf tip (WCT) and leaf base (WCB). d. Temperature difference between the tip (TDT) and the leaf base (TDB) of a glossy privet leaf.

4 Discussion

In the analysis of the damage to plants/trees that is caused by freezing, a long-debated question is why only the leaf tip and/or the branch top are hurt by the almost evenly ranged cold environment. This question has not been addressed by many previous research studies. Using imaging, we found that the structures of both the leaves and branches of Japanese spindle and glossy privet exhibit a heterogeneous property. The abruptly thinner major veins at the leaf tip and/or the margin area results in their lack of sap warming, which makes these plant parts sensitive to freezing damage. Inversely, the leaf and branch base, which are directly connected to the main stem/vascular system, tend to continuously receive warm sap from temperature-stabilized deep underground water.

Although we sometimes observed the warmer leaf venation system by thermography, a significantly higher temperature was found in the major vein in the thermal images obtained after hand heating. In addition, we found consistent correlations between areas with high water content and areas with a high temperature. This finding suggests that a lack of warm sap from underground causes Japanese spindle trees that are newly transplanted or planted in water-stressed environments to be sensitive to freezing damage. The heterogeneity of the major vein from the leaf base to the tip results in the leaf bend observed in Japanese spindle leaves during severe freezing dehydration (Levitt, 1972); this phenomenon is similar to the bending observed during the dehydration of detached leaves in spring. The huge tension in the water transport system easily snaps the water continuum at the thin part of the leaf during freezing dehydration, which obstructs the warm sap from the tip and margin area. Under water stress environments, the leaf tip and margin area are scorched due to a lack of sap warming. The leaf bending phenomenon observed in both freezing-dehydrated leaves and air-dehydrated leaves of Japanese spindle trees suggests that these processes have a similar mechanism. In addition, only a small number of Japanese spindle trees exhibit the leaf scorch symptoms in a normal environment due to their ability to maintain a high water content.

Hand heating effectively increased the temperature difference between the areas of the leaf with different water contents. As the temperature decreased from almost 20 ℃ to −15 ℃, clearer thermal images were obtained with the thermal camera, which clearly show the snap point of the major vein. Therefore, because the different water contents of different leaf areas result in different specific heats, thermography is an effective tool that can be used to identify the different water and thermal states in leaves after hand heating. Thus, thermography can be potentially used to diagnose leaf scorch symptoms before the appearance of visible symptoms.

References

[1] Chaerle L., Caeneghem W. V., Messens E., Lambers H., Montagu M. V. and Van-Der-Straeten D. Presymptomatic visualization of plant-virus interactions by thermography[J]. Nat. Biotechnol., 1999, 17: 813-816.

[2] Chaerle L. and Van-Der-Straeten D. Imaging techniques and the early detection of Plant Stress[J]. Trends in Plant Science, 2000, 5(11): 495-501.

[3] Clements H. F. Significance of transpiration[J]. Plant Physiology, 1934, 9: 165-172.

[4] Gates D. M. Transpiration and leaf temperature[J]. Ann. Rev. Plant Physiol., 1968, 19: 211-238.

[5] Grant O. M., Chaves, M. M., Jones H. G. Optimizing thermal imaging as a technique for detecting stomatal closure induced by drought stress under greenhouse conditions[J]. Physiologia Plantarum, 2006, 127: 507-518.

[6] Jones H. G. Use of thermography for quantitative studies of spatial and temporal variation of stomatal conductance over leaf surfaces[J]. Plant Cell Environ, 1999, 22: 1043-1055.

[7] Jones H. G., Stoll M., Santos T., Sousa C., Chaves M. M., Grant, O. M. Use of infra-red thermography for monitoring stomatal closure in the field: application to grapevine[J]. Exp. Bot., 2002, 53: 2249-2260.

[8] Jones H. G. , Leinonen L. Thermo imaging for the study of plants water relation[J]. Agric. Meteorol. , 2003, 59(3): 205-217.

[9] Kramer P. J. Water Relation of Plants[M]. New York: Academic Press, 2006, 187-213.

[10] Levitt J. Responses of Plants to Environmental Stresses[M]. New York and London: Academic Press, 1972.

[11] Nilsson H. E. Remote sensing and image analysis in plant pathology[J]. Plant Phathol, 1995, 17: 154-166.

[12] Prytz G. , Futsaether C. M. , Johnsson A. Thermography studies of the spatial and temporal variability in stomatal conductance of *Avena* leaves during stable and oscillatory transpiration[J]. New Phytol. , 2003, 158: 249-258.

[13] Rosenberg N. J. Microclimate: The Biological Environment[M]. New York and London: Jone Wiley & Sons, 1974.

[14] Wang F. , Yamamoto H. , Ibaraki Y. Transpiration surface reduction of kousa dogwood trees during seriously losing water balance[J]. Forestry Res. , 2009a, 20(4): 337-342.

[15] Wang F. , Yamamoto H. , Ibaraki Y. Responses of some landscape trees to the drought and high temperature event during 2006 and 2007 in Yamaguchi, Japan[J]. Forestry Res. , 2009b, 20(3): 254-260.

[16] Wang F. , Yamamoto H. Detecting leaf and twig temperature of some trees by using thermography[J]. Spectroscopy and Spectral Analysis, 2010, 30(4): 2914-2918.

[17] Wang F. , Omasa K. Image measurements of leaf scorches on landscape trees subjected to extreme meteorological event[J]. Eco. Infor. , 2012, 12: 16-22.

(原文发表于 *Ecological Informatics*, 2013, 16: 35-40)

Digital image analysis of different crown shape of *Platycladus orientalis* *

Fei WANG[1], Dejun WU[1], Haruhiko YAMAMOTO[2], XING Shiyan[3], Lipeng ZANG[3]

(1. Shandong Forestry Research Academy, Jinan, China, 250014; 2. Agricultural Faculty, Yamaguchi University, Japan; 3. Forestry College of Shandong Agricultural University, Taian, China)

1 Introduction

The crown is the most important photosynthetic organ of woody plants. The size and shape of the crown are biological and ecological characteristics (Maguire and Kanaskie, 2002; Liu *et al.*, 2009) directly associated with or indicating the health status (Waring, 1980) of the plants and affected by the surrounding environment (Noguchi, 1979; Lawrance, 1939). In the past, the studies investigating the meteorological impaction on the crown shape of trees were not less (Liu *et al.*, 2009; Solberg, 1999; Ma and Wang, 2006; Zerl, 2004). We have recently engaged in the study of the crown shapes of trees after typhoon events (Wang *et al.*, 2008; Wang *et al.*, 2009a) and the asymmetrical crown of some landscape tree species (Wang and Zhang, 2011). While studying the crown shape of *Platycladus orientalis* in Shandong Province, China, we observed special tree crowns in torch and fusi-form shapes, reflecting the withering of partial lower twigs, and the normal tower form resulting from the phototropism of lower twigs. The partial withering of the twigs typically occurs in response to changes in the climate, soil and artificial processes. However, the meteorological extremes, such as strong wind, extremely hot weather and drought, which directly impact the water and energy balance of these trees, are the triggers of the branch wilts from lower to the upper parts, resulting in special crown shapes of different forms. Indeed, withering might be

* Corresponding Author: WANG Fei, Email: wf-126@126.com. Supported by the National Natural Science Funds of China (ID Number: 31170671) and the Project of Science and Technology Development in Shandong, China (ID Number: 2012GNC11107)

the result of their structural embolism (Fahn, 1990) to save the life of the tree at the expense of partial twigs and scale leaves (Tyree and Zimmermann, 2002) in the extreme drought and barren hill site environments. This extreme environment adaptation might be a key to drought endurance and importantly ecological and physiological information of the *Platycladus orientalis*.

Many approaches have been used to study tree crown shapes, including visual evaluation, direct measurement with scales, detection with special instruments, regression estimation and image analysis (Redfern and Boswell, 2004). Among these methods, image analysis is simple, fast and time and money saving (Wang *et al.*, 2009a). Combined with the construction of special meteorological indices, we analyzed the special crown shapes of *Platycladus orientalis* using the digital image measurement to construct a simple and easily measured index properly responding their health status and site conditions. This technique might be for the management of non-commercial forest of *Platycladus orientalis* in Shandong Province, China.

2 Materials and methods

2.1 RGB image capture and analysis of *Platycladus orientalis* tree crowns

The image analysis of crown shape is defined as a sideward study of tree crowns using digital image measurement method. The RGB images obtained in the present study are sideward images of the crown shape, including the entire stem, captured at the ground level with an elevation of less than 10 degrees. In the present study, a Fuji SL305 digital camera was used.

To capture the images, the major target trees were separate individuals located at the southeast, south and southwest slope of the hills, except in the special study concerning the shading slope. The mature trees with diameters of more than 10 cm were the primary targets for image capture, not including the trees under severe intraspecific competition in closed stands with the characteristics of small crowns on top of the tree. The juvenile *Platycladus orientalis* trees with vigorous growth were not imaged in the present study except in the use of special figures, because of their thin and imperfect crowns. Trees with clustered crowns were also excluded, as each individual represents a juvenile body. In addition, dying trees with sparse crowns, small scale leaves and dim leaf colors were also excluded because of the stagnation of crown development in these trees. Moreover, crowns with artificially sheared lower branches and evidence of livestock feeding were also excluded from imaging.

The sample sites, which are provided with wide representation and traffic convenience, were selected beforehand in the Google map system. For example, we have investigated almost all counties containing plantations of *Platycladus orientalis* in Shandong and each selected one or two hill sites near the highway or freeway. The crown images were captured at more than 23 sites in Shandong Province and 2 sites in Jiangsu Province in China. The average number of sampling trees in each site is about 30 individuals or more.

The length tool in ImageTool 3. 0 software was used to measure the shape of the crowns in the image analysis (Fig. 1). Subsequently, the crown shape index we invented recently was calculated in pixels by using Equation (1).

$$CSI = C/H \tag{1}$$

in which, the CSI is the crown shape index, C is the width at the widest part of the sideward of the crown, and H is the height from root collar to the level of the widest part of the crown which responds the moving of the core of crown from low to high. In addition, we have previously demonstrated that no significant difference was found in the process of detecting the CSI s of the same tree in different distance and the CSI index was not affected by the image capture distance, although all of the images were not scaled.

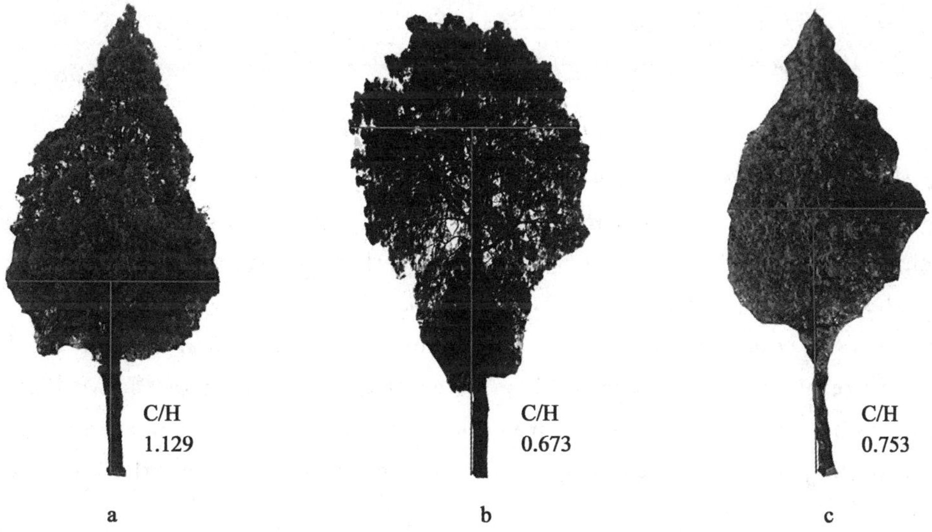

Fig. 1 Typical crown shapes of *Platycladus orientalis* trees.
a. tower form; b. torch form; c. fusi form

After a series of observations, we confirmed that the normal crown shape of the *Platycladus orientalis* is the tower shape (Fig. 1a). Some individuals have exhibited torch (Fig. 1b) or fusi (Fig. 1c) forms in stress environments. In Fig. 1c, the crown shape showed a slight asymmetry, reflecting the impact of persistent prevailing winds. Using the crown shape index (*CSI*), these crown forms were properly determined and analyzed.

In the present study, the color of canopy RGB images has also been analyzed using the RGB analysis method (Wang and Omasa, 2012) accompanying with G/L value calculation (Wang *et al.*, 2009a).

2.2 Construction of Max30twp index, accumulate-arid-index and accumulate-humid-index

In the present study, meteorological analysis was used to confirm the inducer of the abnormal crown shape of *Platycladus orientalis*, in which, the meteorological data were obtained from the China Meteorological Data Service Center. The index was calculated using the meteorological data of daily maximum wind speed, daily precipitation and daily maximum temperature in July, August and September within past 61 years (1951—2011) at 31 meteorological observatory stations in Shandong Province and 10 meteorological observatory stations in neighboring provinces. The detail is described in Equation (2):

$$\mathrm{Max30twp} = \frac{\sum_{m=7}^{9} \mathrm{N}_m \left(\dfrac{\sum_{m=7}^{9} \dfrac{\sum_{n=1}^{30} \max(t_{mn})}{30}}{6} + \sum_{m=7}^{9} \dfrac{\left(\dfrac{\sum_{n=1}^{30} \max(\mathrm{w}_{mn})}{30} \right)}{3} \right)}{\sum_{m=7}^{9} \mathrm{R}_m} - \frac{\sum_{m=7}^{9} \mathrm{P}_m}{180} \tag{2}$$

in which, max30twp is the meteorological index, Max is a function to calculate the maximum value; t_{mn} is the nth daily maximum temperature in $m$month within the 61 years, in which $n=1,2,\ldots,30$; $m=7,8,9$; w_{mn} is the nth daily maximum wind in m month within the 61 years, in which $n=1,2,\ldots,30$; $m=7,8,9$; N_m is the average rain-free-day numbers in m month; R_m is the average rainy-day numbers in m month; and Pm is the mean precipitation in m month. The meteorological relevance of the Max30twp is that the higher the extreme temperature, the wind speeds and the number of mean rain-free days, the higher the index and the larger the mean precipitation, the lower the index.

To analyze the impact on abnormal crown shape, we calculated the accumulative aridity index of ten days (AD10) and the accumulative humidity index of ten days (HD10) using Equations (3) and (4), respectively, for Jinan City in the first half of 2012:

$$AD10_i = \sum_{j=1}^{10} MT_{i+j} / \sum_{j=1}^{10} PR_{i+j} \tag{3}$$

$$HD\,10_i = \sum_{j=1}^{10} PR_{i+j} / \sum_{j=1}^{10} MT_{i+j} \tag{4}$$

in which, $i=1, 2, \ldots, 365$ and $i=1$ on January 1st, $j=1, 2, \ldots, 10$, MT is the daily maximum temperature and PR is the daily precipitation.

2.3 Thermo image capture and wind speed measurement

The thermo images, which were used for measuring the image temperature of *Platycladus orientalis* canopy within and without the valley area, were determined through thermography using a NEC H2640 thermal infrared (8 − 13 μm) camera equipped with a temperature measuring scope ranging from −40−500 ℃ and a minimum sensible temperature of 0.03 ℃. Throughout the measurements, the camera was handheld to capture the vertical thermo images in and out of the valley from the highest point of the nearby hill site. The thermal images were obtained during natural sunny light in the LVT automatic sensitivity-tracing mode and with 0.98 emission efficiency.

The wind speeds in and out of the valley area of the mountain were measured using an AZ8910 wind meter (Hengxin Company, Taiwan) at a height of 1.2 meters and a persistent time interval of 3～5 minutes. The average wind speed value was read from the meter. Subsequently, the seedling numbers were also counted to determine the environmental variance in and out of the valley.

2.4 Data analysis and design of ARCGIS map

The one way ANOVA in Microsoft Excel software 2003 has been used to determine the significant difference among samples with a probability of more than 99% ($p<0.01$) or more than 95% ($p<0.05$) after the Fisher's exact test. Then, statistical graphs were generated using KaleidaGraph 4.0. The ARCGIS map was constructed using Arcgis 9.3 (ESRI Inc.) with both the Max30twp index and crown shape index of *Platycladus orientalis*.

3 Results and analysis

3.1 The difference in the crown shape of *Platycladus orientalis* in the windy mountain areas

Previous studies have indicated that *Platycladus orientalis* is a tree species sensitive to wind damage. However, it is not clear whether this characteristics reflects the lack of concentrated *Platycladus orientalis* plantation in the windy peninsular area of Shandong province. In the prevailing wind tunnel, strong winds often occur in the narrow draughty area between Dazhai Mountain (N36°05′00″, E116°19′9″) and Nantianguan Hill. Although the wind resource is not the highest observed in Shandong Province, the windmills installed in the wind tunnel (Fig. 2a-★) indicate higher wind energy. In late March 2014, the field measurement of the wind speed using a handheld wind meter indeed showed a high wind speed of 10 m/s (Fig. 2a-5) at the wind draughty area, while the wind speed within the valley was only 2 m/s (Fig. 2a-1).

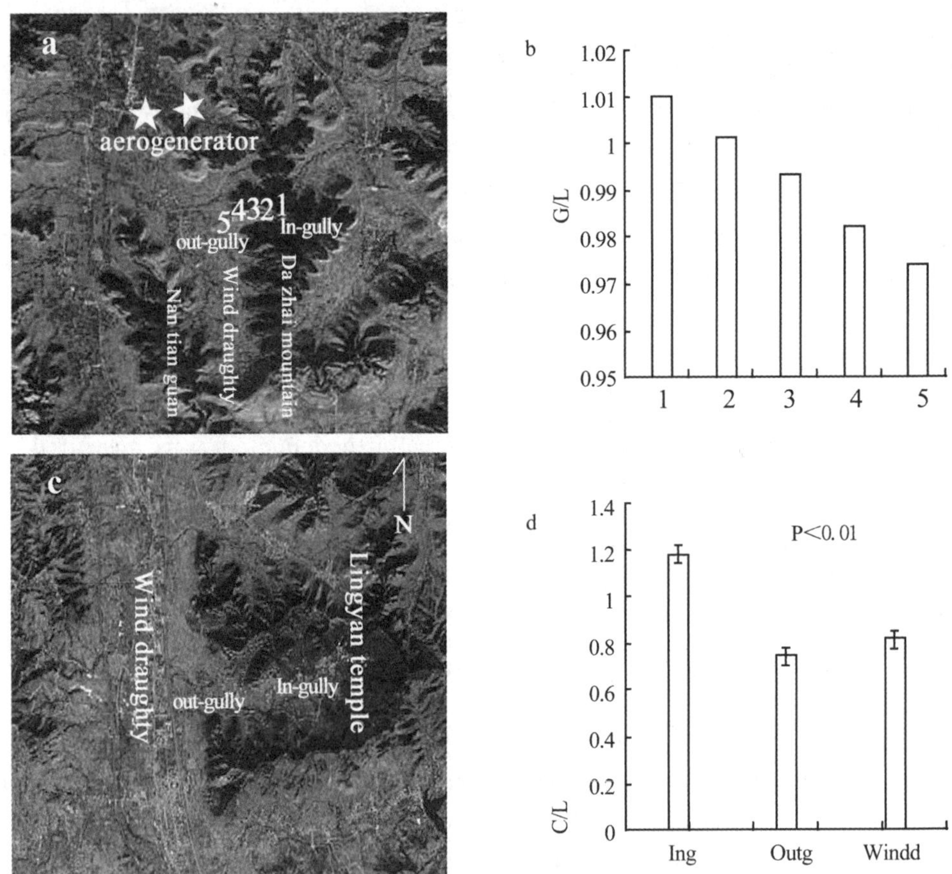

Fig. 2 Environments in a wind draughty area and the difference in crown shape: a. the topology, geography and windy environments near the Dazhai Mountain, in Pingyin, Shandong, China, in which, the symbol ★ stands for the windmills; b. the gradient of G/L value of *Platycladus orientalis* plantation from point 1 to point 5 measured from in-gully to the edge of out-gully at Dazhai Mountain; c. the gully environment near Lingyan Temple, Changqing, Shandong; d. the C/H values for *Platycladus orientalis* trees in the gully (Ing), at the wind draughty area of out gully (Outg) and at the wind draughty area west to Mountain Tai (Windd).

On Dazhai Mountain, *Platycladus orientalis* trees were planted side by side. The individuals in the wind draughty area out-valley showed low stem and sparse canopy in response to the impact of strong winds every year. Some of these trees showed an asymmetric crown (Fig. 1c) in the area within the prevailing wind. The gradually sparse canopy of *Platycladus orientalis* from in-valley to out-valley showed an image gradient from green to gray, and decreasing G/L values were measured (Fig. 2b). Within the wind draughty area near the ridge of Mountain Dazhai, small pole-like *Platycladus orientalis* individuals were also observed on lower hillock site.

Under similar conditions, along a valley of 3—4 km near the Lingyan Temple in Changqing District of Jinan City (Fig. 2c), most of the *Platycladus orientalis* showed tower-shaped crowns within the valley, with larger CSI (Fig. 2d-Ing), although torch-like trees were observed at a local poor site. However, the crown barycenter of most trees gradually rises at the outskirt of the valley and the CSI value of the crowns decreases (Fig. 2d-Outg). Furthermore, many large trees outside the valley to Lingyan Temple and within the wind tunnel around Mountain Tai similarly showed torch or fusi-form crown shapes, with low CSI values (Fig. 2d-Windd).

3.2 Variations of the tree crowns in and out of the valley

Compared with the out-valley north the mesa of Jinan Forest Farm, the crown of mature *Platycladus orientalis* trees located at sunny slopes in the valley primarily showed tower shapes that significantly differed from those outside the valley (Fig. 3a, $p<0.01$). The crown shapes of *Platycladus orientalis* trees outside the valley were primarily torch and fusi-form. No significant crown difference was observed between the mature *Platycladus orientalis* trees located at the shade slope in and out the valley, although there was variation among these individuals (Fig. 3b), likely reflecting their environmental dissimilarity.

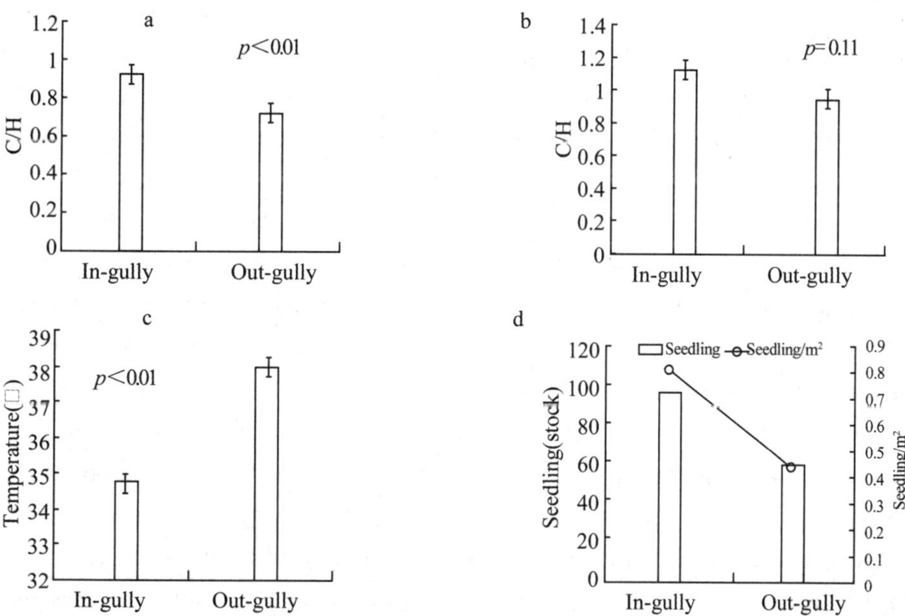

Fig. 3 Differences in the environment, crown shape and plantation status in-gully and out-gully: a. the crown shape difference at Adret; b. the crown shape difference at Ubac; c. the image temperature difference in-gully and out-gully; d. the seedling number difference in-gully and out-gully in a hill site of Jinan Forestry Farm.

In addition, we have measured a significant different image temperatures of *Platycladus orientalis* canopy in the valley and out of the valley by using a thermography (Fig. 3c), which respond to their characteristics of wetness and dryness. In response to the special environment, there were more seedlings or saplings in the valley than out of the valley (Fig. 3d), indicating that the moist valley environment is more suitable for the growth of *Platycladus orientalis*, because their seedlings were more rigorous in adapting to the environment.

3.3 Crown shape and meteorological analysis

In the wind draughty area of the west circuitous flow around Mountain Tai (MT), many low and residual calcareous rock hillocks were observed. This area is located at the rainless area behind MT and the hillside area with lowest rainfall in Shandong Province. The minus budget of water vapor is typically observed in this area and consistent with the normal torch and fusi-form crown shapes of *Platycladus orientalis*. Layering the Max30twp and CSI of *Platycladus orientalis* into the Arcmap (Fig. 4a) revealed that the highest Max30twp index was observed around the wind draughty area of the west circuitous flow around Mountain Tai and along the coastal area where almost no concentrated *Platycladus orientalis* plantation exists.

Most investigated sites around the wind draughty area of the west circuitous flow around Mountain Tai show abnormal crown shapes for *Platycladus orientalis*, with lower CSI, lower precipitation between 600—650 mm during the period over 0 ℃ (Fig. 4a) and green and yellow marks on the map, including sites 10 (Fig. 4a-10), 11 (Fig. 4a-11), 12 (Fig. 4a-12), 13 (Fig. 4a-13), 14 (Fig. 4a-14) and 15 (Fig. 4a-15). In addition, the sites, including sites 9 (Fig. 4a-9), 18 (Fig. 4a-18), 19 (Fig. 4a-19), 20 (Fig. 4a-20) and 21 (Fig. 4a-21), investigated at areas with precipitation lower than 650 mm also showed lower CSI values. With a lower Max30twp index and higher annual rainfall, sites 1 (Fig. 4a-1), 2 (Fig. 4a-2) and 3 (Fig. 4a-3) showed larger crown shape indices and red colored marks. The crown shape index of sites 4 (Fig. 4a-4), 5 (Fig. 4a-5) and 6 (Fig. 4a-6) was slightly inferior and consistent with a relatively lower precipitation during the period over 0 ℃. In agreement with intermediate precipitation, the CSI for sites 7 (Fig. 4a-7) and 8 (Fig. 4a-8) showed an intermediate size.

In addition, we selected three sites at similar longitudes from north to south of East China to compare the crown shape index. The crowns in the Xuyi Hill area in Jiangsu Province (N32°37′50″, E118°14′39″) were primarily tower shaped. In this area, the crown shape index was high (Fig. 4b), although many windmills similar to those in the Dazhai area were observed during the investigation period at the measuring site.

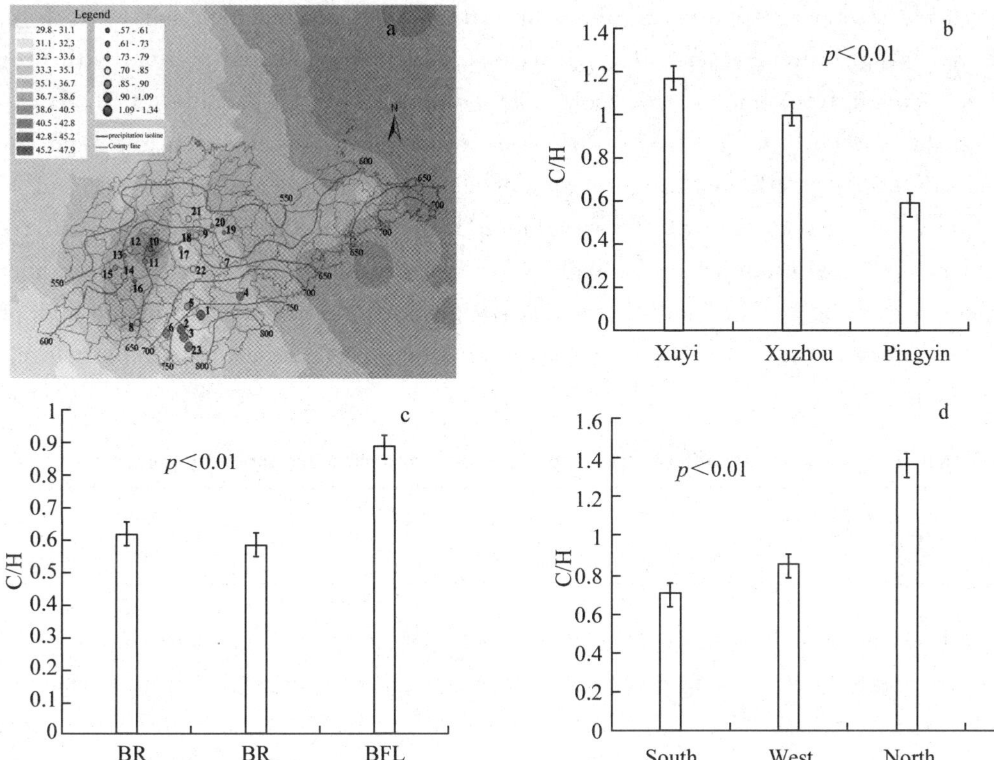

Fig. 4 Crown shape variations of *Platycladus orientalis* trees in different environments: a. the Arcgis map shows the Max30twp index, crown shape index (CSI) and precipitation isoline during the period over 0 ℃ (—) of Shandong Province, China; b. crown shape index (C/H) for Xuyi, Xuzhou and Pingyin; c. crown shape index (C/H) for the trees at sites with different soil types, in which BR is brown rhogosol and BFL is brown forest loam; d. crown shape index (C/H) for the trees at different slopes on the same hill(See attached color Fig. 8).

The CSI for Xuzhou (N34°16′47″, E117°15′18″), Jiangsu at the extension of the mountain chains of Yi and Meng was intermediate among the three sites, whereas the smallest CSI was observed at Pingyin (Fig. 4b), Shandong, located in the northwest. In addition to differences in the CSI, the precipitation among these areas also showed a large difference (Table 1), from 1000 mm at Xuyi to 850 mm at Xuzhou and 658 mm at Pingyin (Table 1). These findings suggest that the crown shape is consistent with the arid and humid degree of the sites, which is indirectly affected by high wind speeds and warmer temperatures that disturb the microclimate of the forest. Indeed, in East China, there is a precipitation gradient from southeast to northwest. However, this tendency is often modified in soil types with different water conservation capacities. In similar precipitation areas, the *Platycladus orientalis* trees in various soil textures often showed different CSI. In Shandong Province, China, most *Platycladus orientalis* plantations are distributed at hill sites with calcareous parent rocks or cinnamon soils, which have relatively high water conservation. However, a little was planted at sites with brown soil, particularly on brown rhogosol. The two sites with the smallest CSI, sites 16 (Fig. 4a-16) and 17 (Fig. 4a-17), and precipitation near 650 mm during the period over 0 ℃ are located in a brown rhogosol area developing from igneous parent rock and possessing lower water and fertile conservation. However, the *Platycladus orientalis* trees at the site 22, with normal brown soil and similar precipitation, show a slightly higher level of CSI (Fig. 4a-22), and the trees at site 23, with normal brown soil and high precipitation (more than 750 mm during the period over 0 ℃), show the highest level of CSI (Fig. 4a-23).

Table 1 Annual precipitation at the sample sites for crown shape measurement (mm)

Pingyin	Xuzhou	Xuyi
658.4	850	1000

These results show large differences in the CSI between trees planted at sites with brown rhogosol, characterized by increased water loss and decreased fertility and trees at brown forest loam (Fig. 4c). These findings reflect increased rainwater runs off from the brown rhogosol compared with the brown forest loam, and the *Platycladus orientalis* trees are also able to endure the droughty environment on brown rhogosol.

In addition, a large difference in the CSI for the trees located at different slopes with different aridity on the same hill site was observed. In the northern slope, most trees presented tower crown shapes and higher CSI (Fig. 4d) in the relative humid environment. Briefly, the meteorological factors, particularly the meteorological events, which affect the crown shape of *Platycladus orientalis* primarily involve the water relations. Furthermore, we have found a significant linear relation between the crown shape index and the precipitation in Shandong Province, China (Fig. 5a). Two nearer *Platycladus orientalis* plantations with both brown forest loam and brown rhogosol soils were even measured significantly different soil water contents as well (Fig. 5b) in spring.

The Yanzishan Forest Station of Shandong Forestry Research Academy is located south of Jinan City and in a semi-arid monsoon climate area. The yearly precipitation in this area showed large variation, although the rainfall occurred with high temperature during the summer. The drought observed during the first half of 2012 might exemplify a typical varied climate. A 22% reduction in the rate of precipitation was observed from January to June 2012, while a 50% reduction was observed from May to June, a 40% reduction was observed in June and a 65% was observed in May. Similarly, the AD10 was frequently observed, while the HD10 was seldom observed (Fig. 6a) in the first half of the year. This environment primarily affected currently planted seedlings at a test field on a hill site, particularly the planting sites with calcareous parent rock. The drought induced the branch dieback and leaf scorch of many Zhonghuahongye poplar (*Populus* × *euramerica* "Zhonghong") seedlings because the growth of these trees favors humid environments. New small leaflets appeared on ginkgo (*Ginkgo biloba* L.) and smoke-tree (*Cotinus coggygria* Scop.), while the needles of the Japanese black pine (*Pinus thunbergii* Parl.) showed the delayed spread. The withering twigs and scale leaves of *Platycladus orientalis* saplings nearby (Fig. 6b) started from lower part of the crown.

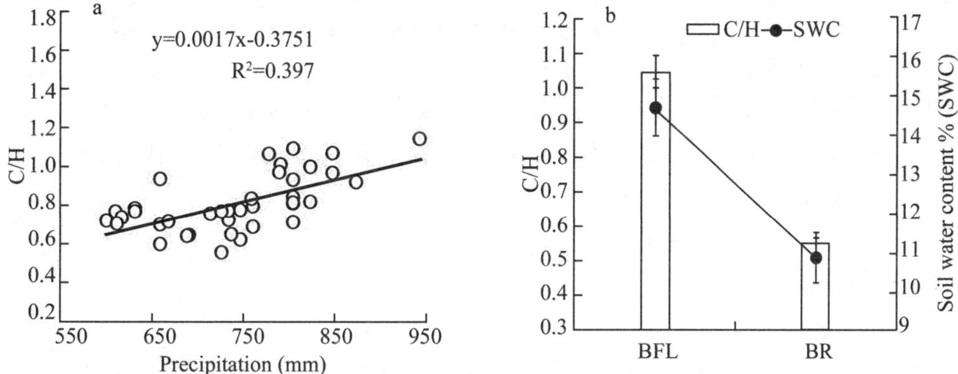

Fig. 5 Relations between crown shape index of *Platycladus orientalis* and water balance: a. linear relation between crown shape index (C/H) and the precipitation in Shandong Province, China; b. crown shape index (C/H) and soil water content for the trees at two sites with brown forest loam (BFL) and brown rhogosol (BR).

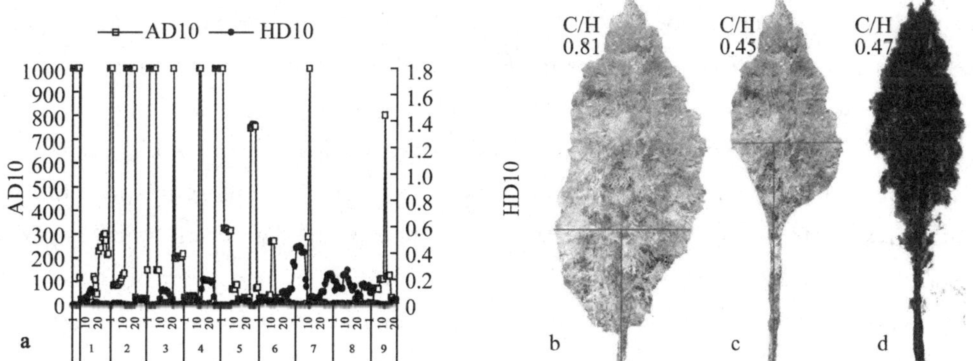

Fig. 6 The drought environment in Jinan, China during the first half of 2012, and the wilt of *Platycladus orientalis* twigs and scale leaves from bottom to the top of the crown: a. the ten days arid index (AD10) and ten days humid index (HD10); b. one example of a sapling crown *of Platycladus orientalis* with significant wilting of the twigs and scale leaves at the lower part of the crown in early summer in 2012 and near a planting test field where many new planting trees showed significant symptoms of drought damage; c. the crown shape after artificially removing the lower wilted part; d. a typical crown shape of young *Platycladus orientalis* trees in a drought rocky hill site.

3.4 Crown shape variance and leaf and branch withering of *Platycladus orientalis*

We examined this phenomenon using image analysis, and the results showed that the crown shape index (CSI) for the residual crown was similar to that of the fusi-form crown (Fig. 6c) when the lower wilted part was artificially removed using image treatment, with a CSI of 0.45, whereas the CSI was 0.81 prior to the removal of the wilted part. In the present study, we observed young trees with typical fusi-form crown shapes and a CSI of 0.47 on barren and rocky hill sites (Fig. 6d). The programmed wilt that changed the crown shape of the *Platycladus orientalis* from tower to fusi-form and torch crown shapes is popular in Shandong Province. The meteorological events, such as drought, strong wind and high temperature as well as winter drying, are the major inducers of these abnormal crown forms of *Platycladus orientalis* trees planted under drought and barren site conditions.

4 Discussion

Trees adapt to natural environments in various patterns (Zierl, 2004). After meteorological events, such as strong wind and dry and hot weather as well as winter drying, some tree species wither (Goerg and Vollenweider, 2007; Kozlowski, 1976) or shed their leaves or twigs (Kozlowski, 1973; Wang *et al.*, 2009b) and show asymmetric crowns (Wang and Zhang, 2011) and wilt of the aboveground structures. These effects are similar to the inverse process of tree growth, and the proper example is the reduction in transpiration surface area (Wang *et al.*, 2009b; Orshan, 1954).

As the major composition of the warm coniferous forest in North China, *Platycladus orientalis* is a pioneer tree species in afforestation at calcareous rock hill sites, showing tolerance to drought and barren soil. These trees are considered irreplaceable tree species planted at calcareous mountain areas. Because of the significant apical superiority of these trees, leaf and branch withering often occurs from bottom of the crowns. High resistance to stress and the gradual and programmed wilt from the bottom to the top induces the torch or fusi form shapes of the crowns of *Platycladus orientalis* trees. These crown shapes are typically associated with the productivity and vigor status of these trees. The crown shape restores to tower form from torch or fusi forms in proper environments when the trees are healthy and undergoing vigorous growth. Similarly, the crowns of *Platycladus orientalis* trees are also able to restore from gnawing and resecting process. In contrast, difficulties in crown restoration are observed when the growth of these trees declines or is slow.

Meteorological factors not only affect the water metabolism of *Platycladus orientalis* but also impact the crown shapes, particularly for individuals planted under poor site conditions. The results of the present study suggest that the strong wind and extreme dry and hot environments are foundational conditions that induce torch and fusiform crown shapes. The masses of *Platycladus orientalis* trees with torch or fusi form crown at extended and remnant hillocks of Mountain Tai in North West Shandong Province are associated with the semi-arid and windy environment. A smaller CSI is often observed in this area, while a larger CSI is observed in southern areas and at superior sites with abundant precipitation. These observations suggest that *Platycladus orientalis* grows well in proper moisture and fertile environments. The proper survey result might be the precondition for the tending and management of classes of crown shapes.

Historically, there are several studies concerning crown analyses (Wang *et al.*, 2014; Liu *et al.*, 2009). However, in the present study, we focused on the sideward crown shapes of *Platycladus orientalis* using digital image analysis. The normal measurement of tree crowns involves time taking, high labor costs and strenuous tasks. In comparison, digital image capture on the ground and relative analysis have the advantages of agility and convenience, as these techniques can be used to quantitatively measure crown shape, and the present study invented a proper indicator of the health status and site conditions of *Platycladus orientalis* plantations. Thus, it will be important to analyze the healthy status of *Platycladus orientalis*, providing a far-reaching impact on site selection for tending and managing non-commercial *Platycladus orientalis* forests.

References

[1] Fahn A. Plant Anatomy[M]. Oxford: Pergamon Press, 1990.

[2] Gnnthardt-Goerg M. S., Vollenweider P. Linking stress wilt macroscopic and microscopic leaf response in trees, new diagnostic perspectives[J]. Environ Pollution, 2007, 147: 467-488.

[3] Kozlowski T. T. Shedding of Plant Parts[M]. New York: Academic Press, 1973.

[4] Kozlowski T. T. Water Supply and Leaf Shedding, in Water Deficits and Plant Growth[M]. New York: Academic Press, 1976.

[5] Lawrance D. Some feature of the vegetation of the Columbia river Gorge with special reference to asymmetric to forest trees[J]. Ecologic Monograph, 1939, 9: 217-257.

[6] Liu P., Ma F. Y., Jia L. M., Wang Y. T. Model of crown composition index and health appraisal of *Pinus tabulaeformis* and *Platycladus orientalis* plantations[J]. Journal of Northeast Forestry University, 2009, 37(3): 32-35. (In Chinese)

[7] Ma Lu-yi, Wang Xi-qun. Growth space competition index (GSCI) and application in the individual intraspecies competition of *Pinus tabulaeformis* and *Platycladus orientalis* forests[J]. Ecological Science, 2006, 25(5): 385-389. (In Chinese)

[8] Maguire D. A., Kanaskie A. The ratio of live crown length to sapwood area as a measure of crown sparseness[J]. Forest Science, 2002, 48(1): 93-100.

[9] Noguchi Y. Deformation of trees in Hawaii and its relation to wind[J]. Journal of Ecology, 1979, 67: 611-618.

[10] Orshan G. Surface reduction and its significance as a hydrocological dactor[J]. Journal of Ecology, 1954,

42:442-444.

[11] Redfern D. B., Boswell R. C. Assessment of crown condition in forest trees: Comparison of methods, sources of variation and observer bias[J]. Forest Ecology and Management, 2004, 188: 149-160.

[12] Solberg S. Crown density assessments, control surveys and reproducibility[J]. Environmental Monitoring and Assessment, 1999, 56: 75-86.

[13] Tyree M. T., Zimmermann M. H. Xylem Structure and the Ascent of Sap[M]. Berlin: Springer-Verlag, 2002.

[14] Wang F., Yamamoto H., Iwaya K. Quantitative research on vigor of ginkgo trees hit by Typhoon 0613 with ground-based digital image analysis[J]. Journal of Natural Disaster Science, 2008, 30(1): 45-53.

[15] Wang F., Yamamoto H., Ibaraki Y., Iwaya K., Takayama N. Evaluating ginkgo leaf necrosis and asymmetric crown discoloration induced by Typhoon 0613 with RGB image analysis[J]. Journal of Agricultural Meteorology, 2009a, 45(1): 27-37.

[16] Wang F., Yamamoto H., Ibaraki Y. Transpiration surface reduction of kousa dogwood trees during seriously losing water balance[J]. Journal of Forestry Research, 2009b, 20(4): 337-342.

[17] Wang F., Zhang J. Q. Asymmetric response of landscape trees to disastrous weathers[J]. Journal of Catastrophology, 2011, 26(2): 5-10,30. (In Chinese)

[18] Wang F., Omasa K. Image measurements of leaf scorches on landscape trees subjected to extreme meteorological event[J]. Ecological Informatics, 2012, 12(2012): 16-22.

[19] Wang Y., Wang D. H., Zhang J. N., Yu Z. Q., Liu J. L. Difference of crown two-dimensional feature of *Platycladus orientalis* plantation under different stand densities in loess area[J]. Journal of Northwest Forestry University, 2014, 29(3): 125-128. (In Chinese)

[20] Waring R. H., Thies W. G., Muscato D. Forest Scicence, 1980,26:112-117.

[21] Zierl B. A simulation study to analyze the relations between crown condition and drought in Switzerland[J]. Forest Ecology and Management, 2004, 188: 25-38.

(原文发表于 *Ecological Informatics*, 2016, 34: 146-152)

Measuring leaf necrosis and chlorosis of bamboo induced by Typhoon 0613 with RGB image analysis*

WANG Fei[1], Haruhiko Yamamoto[2], Yasuomi Ibaraki[2]

(1. The United Graduate School of Agriculture Science, Tottori University, Japan; 2. Faulty of Agriculture, Yamaguchi University, Japan)

1 Introduction

As response to unfavorably extreme environmental stresses such as salt damage, water stress, hot injury, nutrition deficiency, excessive pesticide etc., which are also called physiopath (Treshow, 1970), many plants show symptoms of leaf necrosis or chlorosis at leaf tip and margin even overall leaf. Salisbury (1805) noted that strong typhoon without rainfall often resulted in great leaf injury. The typhoon with less rainfall in Yamaguchi revealed that the leaf necrosis and chlorosis induced by T0613 occurred as Salisbury's note. Following the T0613's hit, leaf necrosis has appeared on many trees and shrubs in Yamaguchi city, especially on bamboo leaves. Leaves on the same crown or canopy appeared not only chlorosis but also necrosis for a period of time.

As an important approach, image color system analysis, especially RGB image analysis, and the concerned applications in measuring plant chlorophyll, nitrogen and disease status have been much reported (Kawashima *et al.*, 1998; Xu *et al.*, 2002; Okado *et al.*, 1993; Suzuki, 1995; Lei *et al.*, 2004). In the research of wheat and rye, Kawashima *et al.* (1998) considered the normalized difference (red − blue)/(red + blue) was the most applicable function using data collected under different meteorological conditions by portable video camera. Iwaya (2005) measured water content of wheat panicle by (R−G)/(R+G) value. Suzuki *et al.* (1999) used the G/(R+G+B) for broccoli

* Received: 2007-12-11; Accepted: 2008-04-03. The online version is available at http://www.springerlink.com. Biography: Wang Fei, researcher in Shandong Forestry Research Institute, China, mainly engaged in the study in the filed of ecophysiology of trees and shrubs. Email: wf-126@126.com. Responsible editor: Hu Yanbo.

identification. Cai (2006) found that R/(R+G+B) was the best parameter for estimating leaf chlorophyll content, and G/R and R/(R+G+B) for estimating carotenoid content of cucumber leaf. In the research of wheat senescence, Adamsen (1999) stated that the relationships between G/R and SPAD were linear over most of the range of G/R and it responded to both chlorophyll concentrations in the leaves and the number of leaves present. The image color analysis has also been applied in the studies about nutrition deficiency of plants (Honami *et al.*, 1992; Xu *et al.*, 2002) and flower number detection (Adamsen, 2000). Although the observation condition, objective plant material and selected optimum indices are different among the researches, almost all of them analyze the RGB image by ratio values and are focused on the color change of the plant, especially on chlorophyll evaluation of the plant canopy (Cai, 2006). Almost no study has been found on the research of leaf necrosis of trees by RGB image analysis, especially on the leaf necrotic area analysis. In this paper, the RGB image analysis has been used in the study not only for leaf chlorosis but also for leaf necrosis of bamboo hit by T0613 with comparing of the G/R and G/L values.

2 Materials and methods

The meteorological data were obtained from Automated Meteorological Data Acquisition System (AMeDAS) of Japan and Hazard Protection Information System in Yamaguchi Prefecture (HPISYP). It includes the maximum wind speed, daily precipitation during the hit of T0613 and monthly precipitation in Sep. and Oct., precipitation during hit by strong typhoon (maximum wind speed over 15 m/s) and 44 days' precipitation after hit by typhoon in the past 40 years (from 1967 to 2006).

52 and 50 bamboo leaves were typically sampled from a bamboo stand in Yamaguchi University for image taking of chlorosis and necrosis, respectively. The images for RGB analysis were taken at a position of 50 cm above the sampled bamboo leaves at natural light condition with a camera (Nicon D70S) mounted on a tripod. The images were stored in the form of JPEG. The resolution of images was 300 dpi and 3000×2000 pixels. According to the known researches (Adamsen, 1999; Iwaya, 2005; Cai, 2006) and the screening of indices in this research, the G/R and G/L values were selected for RGB image analysis. The Red (R), Green (G) and Luminance (L) values were read from the average histogram value of leaves selected by Magic Wand Tool of Photoshop. Since the parts out of the leaves are removed by hand treating, the impact of image noise can be very small by this image process method.

The images for the calculation of NAP were obtained from a Canon scanner (d125u2). NAP is the proportion of necrotic part to overall leaf area (Fig. 1), which is measured by respectively getting pixels of overall leaf and the green part of it. It is calculated by Function (1).

$$NAP = 100 - \frac{pixels\ of\ green\ parts}{pixels\ of\ overall\ leaf} \times 100 \quad (1)$$

Overall leaf　Green part

Fig. 1　Images for green part and overall individual leaves

The images for comparison of G/R value and G/L value were taken from different light conditions, both indoor and outdoor, from early morning to afternoon, with the same camera and image taking method mentioned above. The shutters of camera were respectively 1/1000, 1/50, 1/30, 1/4, 1/3 and 1 for different image taking conditions. SPAD values of individual leaves with chlorosis were impartially measured by using SPAD-502 chlorophyll meter with 30 duplications per leaf, in order to study the relationship between mean SPAD value and G/R or G/L value.

The most seriously damaged bamboo stands in each investigated area were selected from windward of mountains for the comparison of G/R and G/L. The image of bamboo canopy was taken by using a common digital camera (Canon IXY 6.0) about 30～50 m away from the bamboo stands. It is characterized by horizontally taking the image on ground level. The preparing method of image was similar to the method for images of individual leaves. The Distance from Coastline (DC) of every bamboo stands was measured by tool of electronic atlas by the name of E-Atlas Z Professional 5. It was the shortest distance from stand sites to the coastline.

3　Results and analysis

3.1　The characteristics of T0613 and response from trees and shrubs

The T0613 was characterized with lower precipitation, higher wind speed with the maximum wind velocity 20 m/s and daily precipitation 24 mm when it passed through the Yamaguchi City. This period of rainless or less rainfall persisted for more than one month after its hit. According to the data from AMeDAS of Japan, it made a minimum precipitation record in 44 days after strong typhoon's hit, only 8.5 mm. It created a sufficient condition for landscape trees to develop symptoms of leaf necrosis or chlorosis (Table 1).

Table 1 **Related meteorological data for Yamaguchi**

	During Typhoon 0613	Past 40 years		
		Max	Mean	Min
Max. gusty wind (m/s)	42.4	53.1#	39.5#	33.2#
Max. wind speed (m/s)	20.0	28.8*	21.0*	15.4*
Precipitation in the day typhoon hit (mm)	24.0	247.0*	91.2*	5.0*
Precipitation in 44 days after typhoon hit (mm)	8.5	544*	233.1*	8.5*

* The data came from typhoon's hit that maximum wind speed was over 15 m/s.

\# The data came from typhoon's hit that maximum gusty wind speed was over 33 m/s.

Strong typhoon is one kind of disaster that can make serious mechanical damage to trees and shrubs, such as stem breaking, uprooting, bending, slanting and so on. During the hit by T0613, although such kind of damage less occurred, it did injure many landscape trees and induce leaf necrosis or chlorosis at the tip and margin of trees in Yamaguchi City, such as Metasequoia (*Metasequoia glyptostroboides*), Ginkgo (*Ginkgo biloba.*), Trident maple (*Acer Buergerianum*), Blue Japanese Oak (*Quercus glauca*) and Camphor tree (*Cinnamomum camphora*) etc. It made the crown of some landscape trees significant difference between windward and leeward and the injured crowns even can be clearly divided into green part and non-green part (Fig. 2). Leaves in the bamboo stands were discolored after hit by T0613.

Fig. 2 Necrotic leaves and damaged crown induced by T0613

3.2 Measuring leaf chlorosis and necrosis by image G/R values for individual bamboo leaves

Based on the mechanism of SPAD value measurement, it is a sensitive method to the chlorophyll of plant leaves, especially the paddy rice etc. But one measuring data of SPAD value measured by SPAD-502 only respond to the leaf chlorophyll status of 6 mm^2. To perfectly respond the chlorophyll status of overall leaf, a large number of random measurements must be taken to reduce variability and make statistical data comparable. In the study, one sampled leaf was measured with 30 duplications, which was the maximum memory number of SPAD-502 chlorophyll meter. Therefore, it should be the proper estimating value of the chlorophyll status of sampled leaves. However, the G/R value from RGB image analysis in this research was characterized by measuring the overall leaf fast and easily.

Fig. 3 shows a positive linear relationship between G/R value and SPAD value, with $R^2=0.961$. The G/R value ranges from 0.7 to 1.3 and the SPAD value from near 0 to 43. The related function is $Y=-64.14+83.578X$. Higher relationship between SPAD value and G/R value of RGB image implies that the G/R value of bamboo individual leaves can be recognized as a way responding to leaf chlorosis status.

In the research, the NAP was considered as the criterion of necrosis status of individual leaves. The relation between G/R value and NAP of bamboo leaves showed an inverse logistic relationship, with function $Y=100/(1+3.323785E-12\ e^{21.65X})$, $R^2=0.958$ (Fig. 4). It means that as the NAP increases the G/R value deceases smoothly, then sharply and then becomes stable. The relation function seems that the G/R values of leaves with the same NAP of 0% or 100% are not unique. After removing the leaf samples with NAP of 0% and 100%, the relation function almost changed to a linear function, $Y=386.3-362.77X$ (Fig. 5) and the corresponding correlation coefficient $R^2=0.8945$. From this point of view, both chlorosis and necrosis can be quantitatively determined by using the G/R value of RGB image, especially the NAP, which is disable for SPAD measurement and visual estimation.

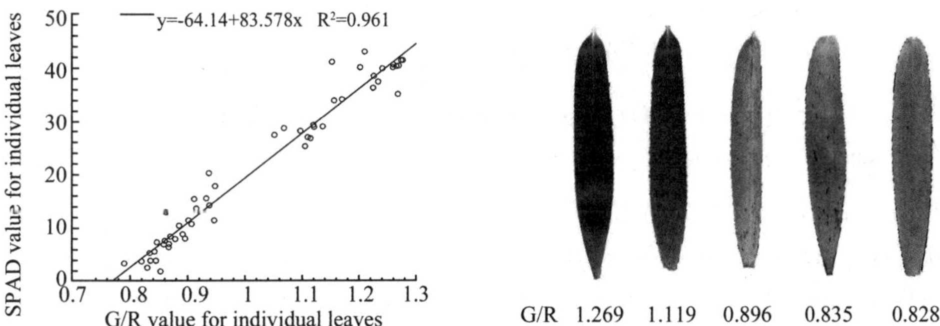

Fig. 3 Relation between SPAD and G/R for individual leaves with chlorosis(leaf figure), and the typical images of bamboo leaf blade with different chlorosis(right figure). In the common situations, bamboo leaves usually existed small necrosis leaf tip. In order to measure the G/R or G/L value of chlorosis leaves, the leaf tip was cut off before image analysis in the research.

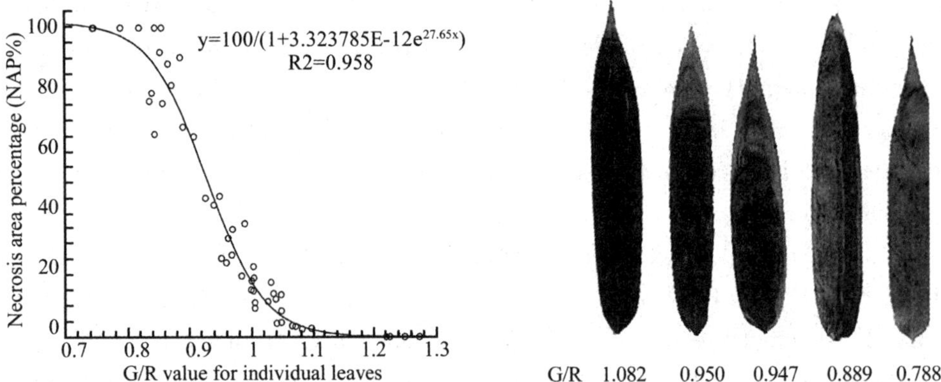

Fig. 4 Relation between G/R value and Necrotic Area Percentage (NAP) for individual leaves(leaf figure), and typical image of bamboo leaf blades with different necrosis(right figure).

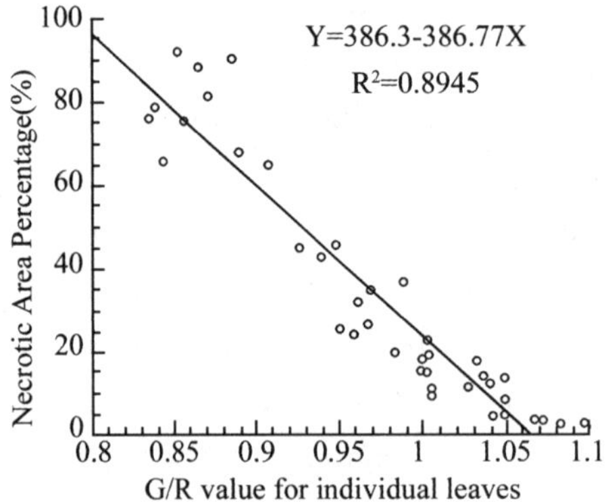

Fig. 5 Relation between G/R value and Necrotic Area Percentage (NAP) after removing the leaves with NAP equaling to 0% and 100%.

3.3 The comparison of G/R value and G/L value of RGB images with big difference in luminance

Kawashima *et al*. (1998) have ever thought that leaf color discrimination with a portable video camera would be difficult under clear conditions with direct solar radiation. Okado *et al*. (1993) considered that the variance of sunlight could make a negative effect on the result of chlorophyll estimation and it could be treated by image correction. We have met the same situation and tried to reduce the effect of light condition by selecting proper indices.

In Lab color system, L value stands for "luminance", which is the linear combination of the R, G and B value from RGB color system. The relationship between L value and R, G, B was showed in Function (2):

$$L=0.299R+0.587G+0.114B \qquad (2)$$

According to this relationship, the R, G and B value vary with the difference of luminance value. According to our test by taking photo image at different light environments for the same leaf, with different L value obtained, it appears bigger variance of G and R value, even the G/R value especially for the green leaves. As the G/R value gets bigger, the difference also becomes larger (Fig. 6A). Theoretically speaking, G/L value decreases with the increasing L value. It may be an index that can reduce the impact of luminance from photo images. The G/L value as shown in Function (3) exists a similar structural character to G/(R+G+B) and R/(R+G+B) as used by other researchers (Suzuki *et al.*, 1995; Cai *et al.*, 2006).

$$G/L=G/(0.299R+0.587G+0.114B) \tag{3}$$

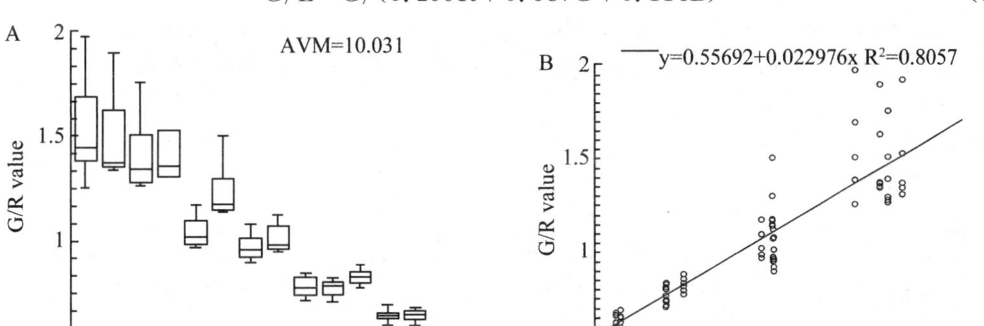

Fig. 6 The Fig. A is the variation of image G/R value for different leaves at different light conditions. The AVM in Fig. A is an average value of variance/mean. The Fig. B is the relationship between G/R and SPAD.

According to the comparison of G/R value and G/L value in the study, the G/L value is more closely related to the SPAD value of sampled leaves from images with big luminance difference than that of G/R value (Fig. 6B, Fig. 7B). It is clear that there is a large difference of variance between G/R value and G/L value for the same sampled leaf at different photo taking conditions. The variance of G/L is about 1/12 of the G/R value and the average variance/mean (AVM) value of G/R and G/L are 10.03 and 0.798, respectively. It is evident that the variance of G/L value among the different images with big luminance difference is significantly lower than that of G/R value (Fig. 6A, Fig. 7A).

Although the G/L value is not able to improve RGB analysis result for the images taken in a large different light condition to the perfectly same, it can give a nearer corrected value from different images with big difference of luminance.

By the image analysis of bamboo stands, a similar result was obtained. The relationship between G/L value for bamboo canopies and the Distance from Coastline (Fig. 8A) was also much closer than that of the G/R value (Fig. 8B) for the images taken at field sites with big light difference.

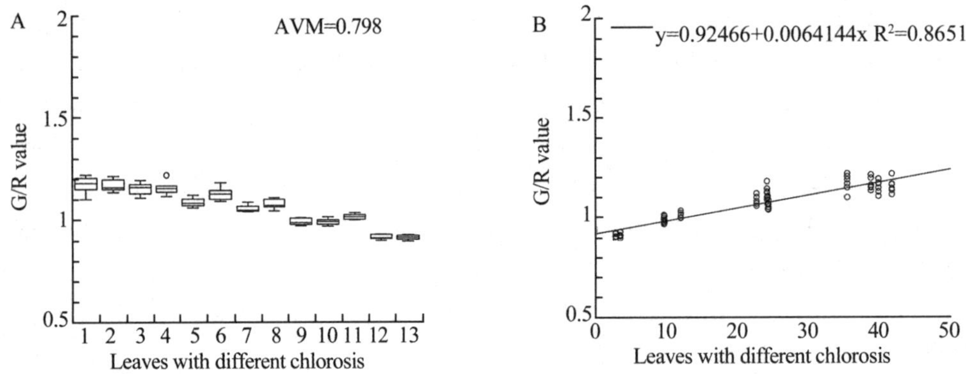

Fig. 7 The Fig. A is the variation of image G/L value for different leaves at different light conditions. The AVM in Fig. A is an average value of variance/mean. The Fig. B is the relationship between G/L and SPAD value for these leaves

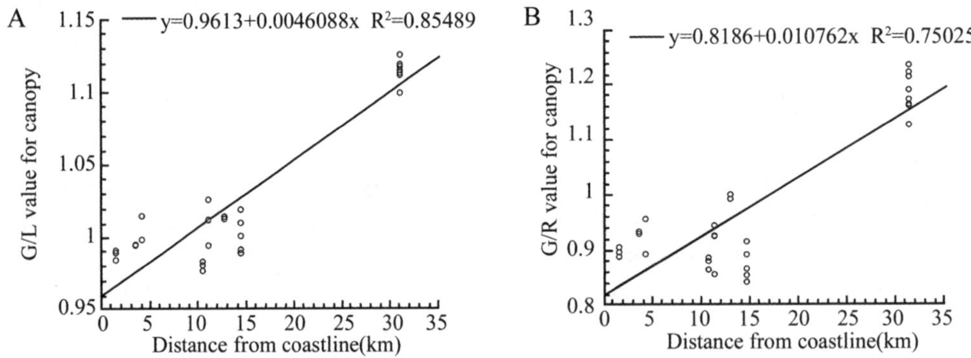

Fig. 8 The Fig. A is the relationship between image G/L value for bamboo canopies and the Distance from Coastline (DC). The Fig. B is the relationship between image G/R value for bamboo canopies and the Distance from Coastline (DC)

However, based on our research, the G/R value can be much closely related to SPAD value of leaves for the RGB image taken at the small light difference conditions, such as scatter light condition of cloudy day or indoor natural light conditions. For example, Fig. 3 showed a close relationship between G/R value and SPAD value. It was the result coming from three images taken at indoor scatter light condition. The concerned correlation coefficient for G/L value was 0. 843, with regression function $Y = 0.0032X + 0.985$. At this kind of condition, G/L value maybe goes beyond the proper limits in correcting the big luminance difference.

4 Conclusion

The strong typhoon without rainfall associated can result in not only mechanical damage to the trees and shrubs, but also physiological injury. The hit by T0613 with less rainfall becomes an example that made many trees and shrubs appear leaf necrosis or chlorosis at leaf tip and margin in a large area. Based on the RGB image analysis, not only chlorosis but necrosis also can be quantitatively determined by measuring image G/R or G/L value of bamboo leaves. It appears a positive linear relationship between G/R value and chlorosis of bamboo individual leaves, and a significant inverse logistic relationship between G/R value and necrosis of bamboo individual leaves for indoor taking images. It has a potential to be an alternative way to determine the vigor status of bamboo leaves damaged by typhoons like T0613.

Almost no research has been found to compare G/R value and G/L value. Based on the present research, the G/L value can get a closer relation with the SPAD value of sampled leaves for the RGB image taken at conditions with bigger luminance difference, and the variance of G/L value is lower than that of G/R value, especially for green leaves. It indicates that the relationship between G/L value for bamboo canopies and the Distance from Coastline can also be much closer than that of the G/R value for the images taken at field sites with big light difference. The G/R is more suitable to be used to analyze the RGB image taken at the conditions with small light difference.

Compared with traditional visual scale method, the RGB image analysis provides a simple and fast tool to estimate the leaf necrosis and chlorosis hit by typhoons like T0613. It will make it possible to conduct a mass investigation in a large scale of area for its less labor need and less time consumption.

5 Acknowledgements

We express our gratitude to Research Lab. of Environmental Ecology, Faulty of Agriculture of Yamaguchi University for providing research condition and thanks to all members who help our field studies.

Reference

[1] Adamsen F. J., Pinter P. J. Jr., Barnes E. M., *et al*. Measuring wheat senescence with a digital camera [J]. Crop Sic, 1999, 39: 719-724.

[2] Adamsen F. J., Coffelt T. A., Nelson M., *et al*. Method for using images from a color digital camera to estimate flower number[J]. Crop Sci, 2000, 40(3): 704-770.

[3] Cai Hongchang, Cui Haixin, Song Weitang, Gao Lihong. Preliminary study on photosynthetic pigment content and color feature of cucumber initial blooms[J]. Transactions of the CSAE, 2006, 22(9): 34-38. (in Chinese)

[4] Iwaya K, Yamamoto H. The diagnosis of optimal harvesting time of rice using digital imaging[J]. Journal of Agricultural Meteorology, 2005, 60(5): 981-984.

[5] Kawashima S., Nakatani M. An algorithm for estimating chlorophyll content in leaves using a video camera [J]. Annals of Botany, 1998, 81: 49-54.

[6] Lei Yongwen, Wei Changzhou, Ye Jun, Hou Zhenan, Li Junhua, Jia Liangliang. Application of computer aided cotton leaf color analysis in nitrogen status diagnosis in cotton plants[J]. Journal of Shihezi University (Natural

Science), 2004, 22(2): 113-116. (in Chinese)

[7] Okado M., Nakamura Y. Studies on the measurement of the color of rice leaves by image processing[J]. Journal of Agriculture Mechanical Association, 1993, 55(5): 75-81. (in Japanese)

[8] Salisbury R. An account of a storm of salt, Linn. Soc[J]. Landon Trans, 1805, 8: 286-290.

[9] Suzuki T. Measurement of growth of plug seedlings by image processing in broccoli[J]. Acta Horticulturae, 1995, 399: 333-343.

[10] Suzuki T., Murase H., Honami N. Non-destructive growth measurement cabbage pug seedlings population by image information[J]. Journal of Agriculture Mechanical Association, 1999, 61(2): 45-51. (in Japanese)

[11] Treshow T. Environment and Plant Response[M]. Mcgraw-Hill Publications in the Agricultural Science, 1970.

[12] Xu Guili, Mao Hanping, Li Pingping. Extracting color features of leaf color images[J]. Transactions of the CSAE, 2002, 18(4): 150-154. (in Chinese)

(原文发表于 *Journal of Forestry Research*, 2008, 19(3): 225-230)

Quantitative research on vigor of ginkgo trees hit by Typhoon 0613 with ground-based digital image analysis

WANG Fei[1], Haruhiko Yamamoto[2], Kiyoshi Iwaya[2]

(1. The United Graduate School of Agriculture Science, Tottori University, Japan; 2. Faculty of Agriculture, Yamaguchi University, Japan)

1 Introduction

Vigor of trees can be thought as one kind of ability to form a perfect individual and healthily grow, which varies with the cultivated conditions, increases with fine nurtures and decreases by serious pest damages, disaster destroys and various kinds of stresses, such as drought, salt, nutrition shortage. Under stressed situations, many unhealthy symptoms may appear like severe wilt, defoliation, discoloration, chlorosis, necrosis, dieback and so on. In some extent, it is a near synonym with tree viability or health. Typhoons are one kind of disaster that can seriously damage forests, trees and shrubs (Hayes, 1999; Yamamoto, 1979; Takahashi and Tani, 1981). The typhoon No. 13 in 2006 (Typhoon 0613) was characterized by low precipitation and high wind speed when it hit Yamaguchi City on Sep. 17th, 2006. Its maximum wind velocity reached 20 m/s, the daily precipitation was only 24 mm during its hit and this period of less rainfall persisted for more than one month after its hit (Table 2). Although there was no severe mechanical damage to trees, such as uprooting, stem breaking, bending, leaning and so on in Yamaguchi City, it did reduce the health status of many tree species, especially some precious landscape trees. Apparent symptoms of leaf discoloration and defoliation appeared on windward of the crowns that made the crown obviously different between the windward and leeward of some deciduous trees, even half green and half brown, such as ginkgo.

Historically, a lot of researches had focused on the storm effect on trees, even making trees as wind indicator, such as the well-known Fujita Tornado Scale and Saffir-Simpson Hurricane Scale, as well as the Griggs-Putnam and Yoshino tree deformation index to predict wind speed and wind direction in meteorological fields (Cullen, 2002; Hennessey, 1980; Wade and Wendell, 1979; Robertson, 1987). Recently, most research works on storm damage to trees still use visual scale method (Okinaka and Sugahara, 1990; Shimizu, 2004; Zhu and Xie, 2001). Visual assessments of tree crown are also common in forest health investigations (Redfern and Boswell, 2004). To some extent, they are observer-specific and probably affected by subjective judgment (Doswell and Burgess, 1988; Solberg, 1999). There is a tendency of transformation to objective methods for determining damage by typhoons and other disasters, especially using imagery analysis nowadays.

Compared with annual plants, trees have big bodies and complex three-dimensional structure, which makes them difficult to be measured. It is almost impossible to perfectly measure tree crowns in a large-scale area for few researchers with common sampling method. As a repaid, nondestructive, noninvasive method, ground-based digital image analysis has been used to measure leaf area index, gap fraction (Bréda, 2003), crop coverage (Purcell, 2000), pest damage and so on. It has been remarked to be a potential method in coverage research of grasses (Richardson *et al.*, 2001), crops (Iwaya, 2003; Purcell, 2000) and vegetation (Laliberte *et al.*, 2006). Richardson *et al.*, (2001) considered that digital image analysis had potential in the study that the amount of green tissue was an indication of health or growth, including injury ratings of various grasses and diseases or insect injury. It has also been used in tree measurement and researches on the porosity of shelterbelts (Kenny, 1987; Guan *et al.*, 2002; Wan *et al.*, 2005). Kenny (1987) concluded that the porosity of shelterbelt could be estimated to be within 2% at a probability level of 0.05 by silhouette method, and the distance of taking photo has no appreciable effect on estimation of porosity. After improvement of photograph treatment method, Guan *et al.* (2002) considered it was a proper way to measure porosity of shelterbelt with high accuracy. In this paper, distinguished from commonly used hemispherical image analysis method, the overall profile of crown was quantitatively measured after hit by Typhoon 0613 by using ground-based, vertical sideward digital images of the profile for ginkgo trees. The vigor status of ginkgo trees hit by Typhoon 0613 was quantitatively studied using indices including the green coverage ratio of crown (GCRC), crown coverage (CC), vigor index (VI) and so on. Combined with analysis of meteorological data, the reason of reducing the vigor status of ginkgo trees has also been studied.

2 Materials and methods

2.1 Research site and basic meteorological data during Typhoon 0613's hit

The research site is located in the area from 131°16′ to 131°45′ east longitude and from 33°55′ to 34°25′ north latitude. The investigation was practiced in a long, narrow area near Yamaguchi Bay, Fushino River and Anno River, which was from seashore via plain to canyon. It includes the circled sites of Ube, Aio, Ogori, Yamaguchi, Miyano, Mitani and Tokusa, which don't match up the administrative area with the same name (Fig. 1), and runs from southwest to northeast.

The meteorological data come from the Automated Meteorological Data Acquisition System (AMeDAS) and from the nearest observation station of the Hazard Protection Information System in Yamaguchi Prefecture (HPISYP). The max wind speed distributes from the highest of 27 m/s in Ube City to the lowest of 8 m/s around Tokusa Town during the hit by Typhoon 0613. The distance to the nearest coastline is from the shortest of less than 1 km to the longest of more than 40 km (Table 1).

Table 1 Basic meteorological data for investigated areas during hit by Typhoon 0613 on Sep. 17th, 2006

	Daily precipitation (mm)	Average wind speed (m/s)	Maximum wind speed (m/s)	Distance from coastline (km)
Aio	18	No data	No data	0.9
Ube	14	10.8	27	1.0
Ogori	21	No data	No data	3.4
Yamaguchi	24	5.1	20	13.4
Miyano	18	No data	No data	19.5
Mitani	19	No data	No data	31.2
Tokusa	38	2.5	8	40.1

2.2 Photo image taking method and standard of analysis

Photographs were taken 45 days after Typhoon 0613's hit. It is characterized by horizontally taking photos of the vertical profile of sample trees on the ground and by using a CCD digital camera (Canon IXY 6.0). The photo-taking distance was determined according to the size of crown to make crown fit the screen of the camera. The position of photo-taking was fixed by moving around the sample tree until the crown could be clearly classified into green part and non-green part and the ratio of the green part to the non-green part didn't change, so that we could obtain the exact sideward photo image of sampled individuals. For trees whose crown cannot be clearly divided into green part and non-green part, the position of photo taking was determined by wind direction, local topography or by reference to other tree species etc. The absolute geographical position of sampled trees was fixed by GPS with Ricoh camera (Caplio 500SE). Photographs were taken between 09:00JST and 16:00JST and at the lightward.

In order to avoid effects by other trees and take photos easily, most of street ginkgo trees were selected in open sites. Trees exhibiting any of the following characteristics were excluded from use in this study to decrease the primary source of error, such as trees sheltered by houses, buildings and other trees, newly planted trees and newly pruned trees as well as trees with scars on the trunk and so on. To avoid errors in analysis, almost all of data used in the results analysis are relative values from the same crown.

2.3 Establishment and measurement of indices

2.3.1 Green coverage ratio of crown

In order to analyze the phenomenon of leaf discoloration of crown and the difference between green part and non-green part on different trees quantitatively, GCRC was applied, which was a proportion of the pixels of green part to the pixels of overall profile of the crown. Firstly, the photo image was treated by image editor software such as Photoshop CS, etc. to remove the parts out of the sampled crown. After obtaining pixels of overall crown, the parts out of green were removed by eraser (Fig. 2) and then pixels of green part of the crown were obtained. The detail of calculation for GCRC by pixel proportion method is presented in Equation (1).

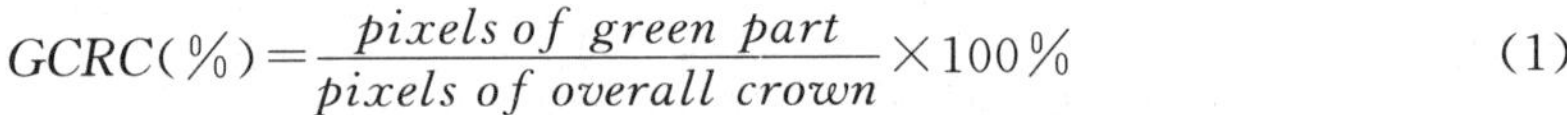

$$GCRC(\%)=\frac{pixels\ of\ green\ part}{pixels\ of\ overall\ crown}\times 100\% \qquad (1)$$

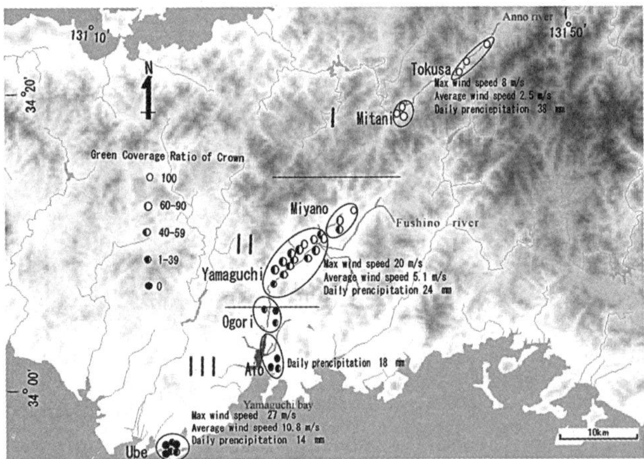

Fig. 1 Integrated map for research area, meteorological data and green coverage ratio of crown (GCRC). Sampled crowns were clearly divided into three groups, Group 1 (Ⅰ), Group 2 (Ⅱ) and Group 3 (Ⅲ).

2.3.2 Crown coverage and vigor index

Another characteristic of ginkgo trees damaged by Typhoon 0613 is defoliation, which is expressed into increasing of openness of the crown. In the study, the openness induced by defoliation of ginkgo trees hit by Typhoon 0613 was estimated by CC index.

It is the pixel proportion of crown silhouette to crown shadow shown in Fig. 3. It was measured by pixel method in reference to researches by others (Kenny, 1987; Guan *et al.*, 2002). The detail procedures of measurement include photo image processing and silhouetting by decreasing the color depth to 2-color palette by using Paint Shop Pro X in the pattern of the blue palette component, the nearest color reducing method and non-palette weight. The pixels of the image shadow were obtained from Photoshop CS. The calculating formula of CC is shown in Equation (2).

$$\text{Crown Coverage}(\%)=\frac{pixels\ of\ silhouette}{pixels\ of\ shadow}\times 100\% \tag{2}$$

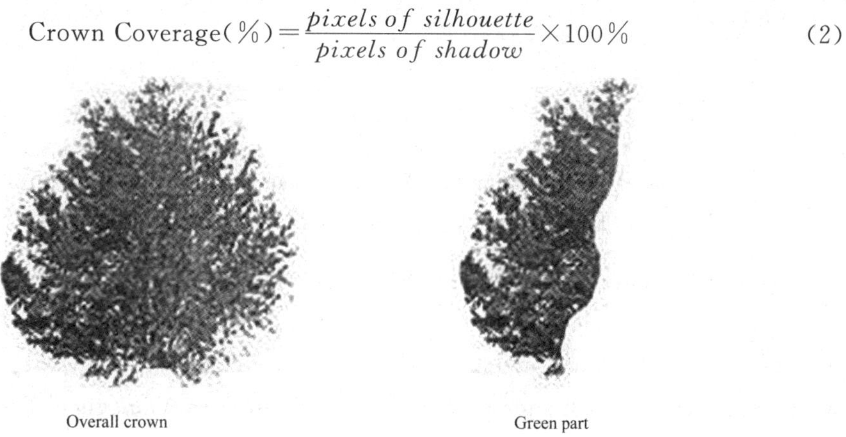

Fig. 2 Damaged crown and green part of ginkgo biloba. Index of Green Coverage Ratio of Crown (GCRC) was calculated by both of them.

Fig. 3 Images of silhouette and shadow for the same crown. Crown Coverage (CC) index was calculated by both of them.

The comprehensive index of VI for damaged trees was calculated by the average value of GCRC and CC, see Equation (3).

$$\text{VI}(\%)=\frac{(\text{GCRC}+\text{CC})}{2} \tag{3}$$

2.3.3 Crown Ratio between Windward and Leeward (CRWL)

As a factor of symmetric characteristics of crowns for multi-analysis, CRWL is the proportion of pixels from the shadows of both windward and leeward of crown divided by reference to the main stem. Firstly, the photo image was processed to remove the parts out of the crown. Then, the crown was divided into windward and leeward from main stem of the tree. After that, both windward and leeward were shadowed respectively and pixels were obtained by using Photoshop CS. The CRWL was calculated as Equation (4).

$$\mathrm{CRWL}(\%)=\frac{pixels\ for\ windward}{pixels\ for\ leeward}\times 100\% \tag{4}$$

2.4 Distance from coastline (DC) and average distance from meteorological station to the coastlines of west, southwest, south and southeast (ADC)

As a main analysis factor, the DC is the shortest distance from tree sites to coastline measured by an electronic atlas named Atlas Z Professional 5.

The average distance from observation stations for AMeDAS to the coastlines of west, southwest, south and southeast was measured in Yamaguchi Prefecture to study the relation between wind speed and the distance from coastline. It was also measured by using Atlas Z Professional 5 and calculated by Equation (5).

$$\mathrm{ADC}=\frac{\frac{west}{2}+\mathrm{southwest}+\mathrm{south}+\mathrm{southeast}}{4} \tag{5}$$

where, west/2 was used for the reason that the average width of the west and east is about 2.3 times more than that of south and north for Yamaguchi Prefecture.

2.5 Multiple statistic analysis

Principle component analysis and cluster analysis were carried out using GCRC, CC, VI, CRWL and DC mentioned above. The analysis was conducted by commonly used software. The distance used in cluster analysis is the square Euclidean distance [Equation (6)] for samples and cluster mean [centroid method and refer to Equation (7)] for classes:

$$d(x_i,x_j)=\left[\sum_{k=1}^{p}(x_{ik}-x_{jk})^2\right]^{\frac{1}{2}} \tag{6}$$

where $d(x_i, x_j)$ is the Euclidean distance between sample i and sample j, $i=1, 2, \ldots, n$, $j=1, 2, \ldots, n$ and $k=1, 2, \ldots, p$. x_{ik} is the data of sample i at point k and x_{jk} is the data of sample j at point k.

$$D_{pq}=d(\bar{x}_p, \bar{x}_q) \tag{7}$$

in which D_{pq} is the centroid distance between class p and class q. $\bar{x}_p$ is the cluster mean value in p class and $\bar{x}_q$ is the cluster mean value in q class.

3 Results and analysis

3.1 Discoloration of ginkgo crowns hit by Typhoon 0613

Based on the calculation, the GCRC distributed from zero or close to zero near the coastline to 100% in the valley far from the coastline around Tokusa Town, corresponding to the overall crown brown and overall crown green. From Fig. 4, although the GCRC values for samples at the same site differed from each other for the reason of different site conditions and growth situations, it was not as great as to significantly effect the relation to DC. A positive logistic function relationship between GCRC and DC was obtained, with a square correlative coefficient of $R^2=0.913$ at 0.01 probability level by screening among the regression equations of logistic, logarithmic, exponential, linear, polynomial etc. The optimal equation determined by maximum correlative coefficient is shown in Equation (8). In addition, the equation was computed from the regression analysis of 36 sampled trees.

$$\text{GCRC}=101.95/(1+8.821e^{-0.183\text{DC}}),\ R^2=0.913 \tag{8}$$

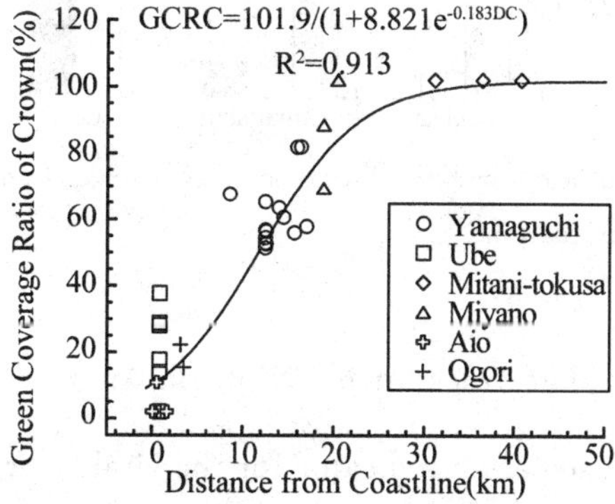

Fig. 4 Relationship between Green Coverage Ratio of Crown (GCRC) and Distance from Coastline (DC)

It was regressed from 36 ginkgo trees from the sites of Yamaguchi, Ube, Mitani-tokusa, Miyano, Aio and Ogori respectively.

It indicates that the further is from the coastline, the greater the green part of the crown is. In other words, the nearer is to the coastline, the more the green leaf color of ginkgo loses. As the distance from coast to inland increases, the GCRC increases sharply, then smoothly, and then becomes stable.

Further, the leaves on a damaged tree were analyzed by dividing them into groups of non-scorched, scorched, and dried leaves by sampling method in order to analyze the pattern of leaf discoloration. Fig. 5 showed the leaf component of samples from trees with different DC. Almost all of the leaf samples from ginkgo trees in Tokusa, with an average DC equaling 35.6 km, were composed of non-scorched leaves, with a non-scorched leaf rate of 93.7%, a scorched leaf rate of 6.3%, and a dried leaf rate of 0%. The scorched leaves showed only scorch spots, which perhaps were not induced by Typhoon 0613 since the scorch spots existed on the leaves at both leeward and windward. Contrarily, the major leaf samples from Ube, with DC equaling 1.7 km, were scorched and dried leaves, with a non-scorched leaf rate of 1.6%, a scorched leaf rate of 49.9%, and a dried leaf rate of 48.5%. The leaf samples from Yamaguchi, with DC equaling 13.5 km, were in the middle position with a non-scorched leaf rate of 39.3%, a scorched leaf rate of 52.3% and a dried leaf rate of 8.4%. So, it can be said that the difference of crowns in different areas mainly comes from different scorch components of leaves 45 days after Typhoon 0613's hit.

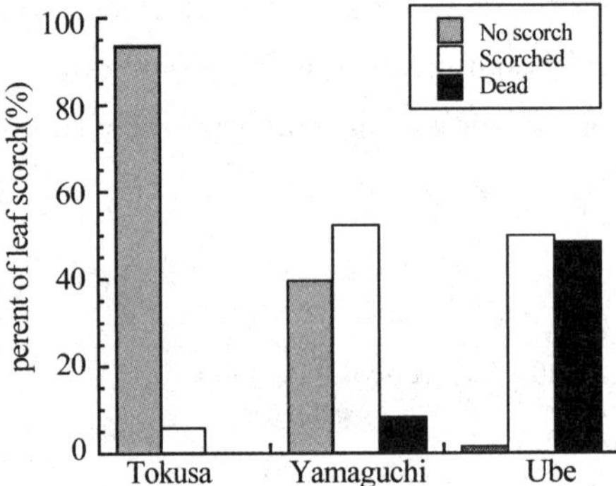

Fig. 5 Percentage of leaf scorch for discolored ginkgo crowns from different sites with different distance from coastline.

3.2 Defoliation of ginkgo trees hit by Typhoon 0613

Defoliation is widely used as an indicator for the vitality or health of forest trees and the damage status in forest investigations, although it is still debatable (Zierl, 2004). It is observed that different defoliation occurred on ginkgo trees in varied site conditions after hit by Typhoon 0613. From Fig. 6, it is evident that CC ranged from 40% to 90% or so and almost no CC value of ginkgo trees became zero or near zero because there were a lot of dead leaves remained on the damaged trees until next spring. Meanwhile, the result appears that there is a positive relationship between CC and DC, although the correlative coefficient is less than that of the relationship between GCRC and DC. The corresponding equation is:

$$CC=101.34/(1+1.0076e^{-0.0535DC}),\ R^2=0.622 \tag{9}$$

The result shows that there is a difference in crown coverage among the sampled trees and the further is from the coastline, the bigger the crown coverage of ginkgo trees is. In other words, defoliation occurred indeed and was more serious near coastline.

The further regression analysis by classifying samples into two groups of dense crowns and sparse crowns showed a positive relationship between CC and DC, and the square correlative coefficients for regressive equations were 0.78 and 0.79 respectively for the dense crown group and sparse crown group. It indicates that the relationship between CC and DC is affected by density of crowns. Therefore, sampled trees with much dense crown were eliminated from the analysis.

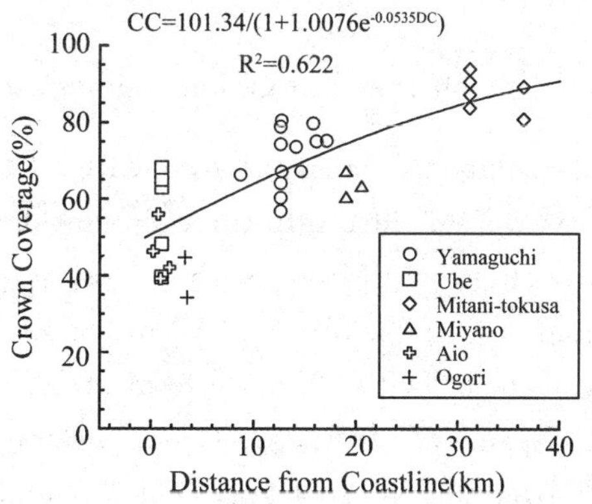

Fig. 6 Relationship between Crown Coverage (CC) and Distance from Coastline (DC). It was also regressed from the same ginkgo trees used in Fig. 3

3.3 Comprehensive vigor status of ginkgo trees hit by Typhoon 0613

Trees' vigor has been evaluated by various methods, which include foliage based indices, volume increment and height growth rate based indices (Robichaud and Methven, 1991). However, the vigor of ginkgo trees to be estimated in this study is the status after hit by strong Typhoon 0613, which is characterized by clear discoloration and defoliation of typhoon damaged trees. Therefore, discoloration and defoliation were used to establish the vigor index to respond to the vigor status of ginkgo trees hit by Typhoon 0613. GCRC and CC are two indices respectively responding to them in some extent and the VI, which integrates indices of GCRC and CC has potential to comprehensively model the vigor status of damaged trees. Fig. 7 gives a relation curve between VI and DC and it shows that there is also a positive relationship between VI and DC, with a square correlative coefficient of $R^2=0.882$ and regressive Equation (10).

$$VI=99.688/(1+2.5366e^{-0.11DC}),\ R^2=0.882 \qquad (10)$$

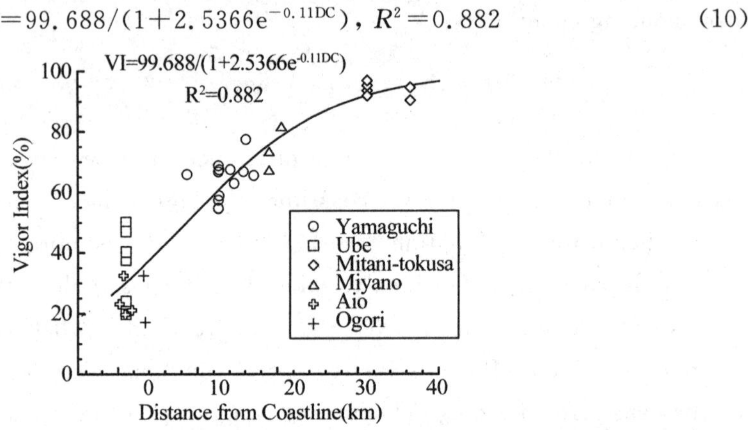

Fig. 7 Relationship between Vigor Index (VI) and Distance from Coastline (DC). Ginkgo crowns analyzed in it were also the same as in Fig. 3.

3.4 Multi analysis and classification of ginkgo vigor status

Based on the principle component analysis by GCRC, CC, VI, CRWL and DC, samples from different areas were divided into three groups showed in Fig. 8. Group 1 consisted of samples from Tokusa and Mitani with DC of more than 30 km, GCRC of 100% or near 100%, average CC of 83.4%, and VI of 95.4%. Group 2 included samples from Miyano and Yamaguchi, with DC from 8km to 20 km, GCRC from 40% to 90%, average CC of 67.3%, and VI of 67.5%. Group 3 came from samples from Ogori, Aio, and Ube with DC from 0.2 km to 4 km, GCRC from 0% to 39%, average CC of 54.2%, and VI of 21.1%.

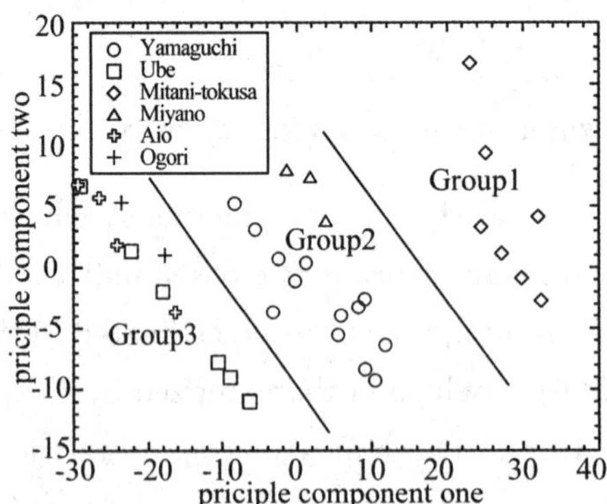

Fig. 8 Classification of ginkgo crowns by principle component analysis. 36 crowns collectively distributed in three areas of principle coordinate system and were classified into three groups. The first group is concentrated in the first and fourth quadrants at the right side of the level axis, the second group around the datum point, and the third group in second and third quadrant at the left side of the level axis.

The result of principle component analysis is evidence that the vigor of ginkgo trees were more seriously damaged by Typhoon 0613 within 4 km from the coastline, and almost no injury occurred in the area out of 20 km from coastline and the ginkgo trees in the area from 8 km to 20 km were in the middle position.

Almost the same result has been obtained by Euclidean distance cluster analysis at the point of squared central distance equaling 6. 05 with the data of GCRC, CC, VI, CRWL and DC according to the discriminating standard of the starting point that the squared central distance sharply increases. The samples from different areas also can be divided into three groups as showed in Fig. 9. The samples from Group 1 consisted of samples from Tokusa and Mitani, Group 2, from Yamaguchi and Miyano and Group 3, from Ube, Aio and Ogori except only one special sample from Yamaguchi.

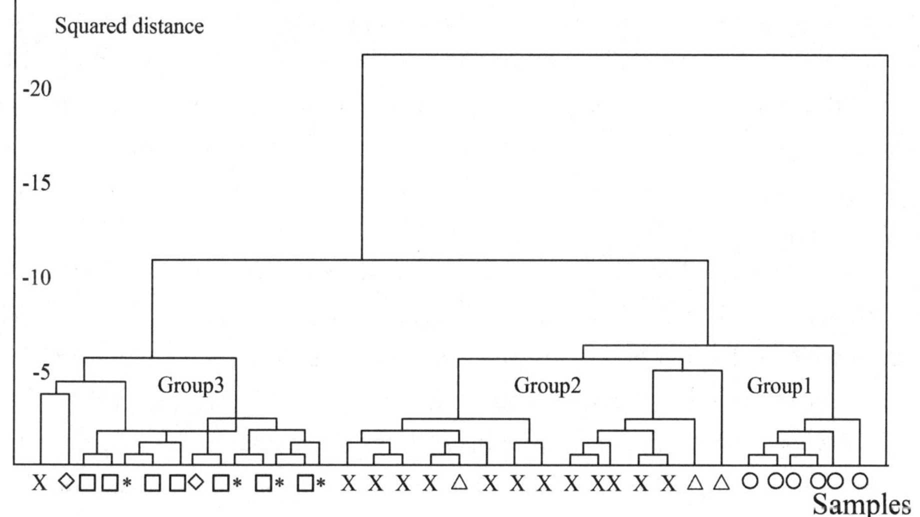

Fig. 9 Result of cluster analysis with centroid method, in which the crowns from Mitani-tokusa (○), Miyano (△), Yamaguchi (X), Ogori (◇), Aio (*) and Ube (□), were also classified into three groups.

Fig. 10 shows a few of models of ground-based digital image samples for Group 1, Group 2 and Group 3 respectively. A great difference among the groups was shown and they were consistent with the indices used in this research properly.

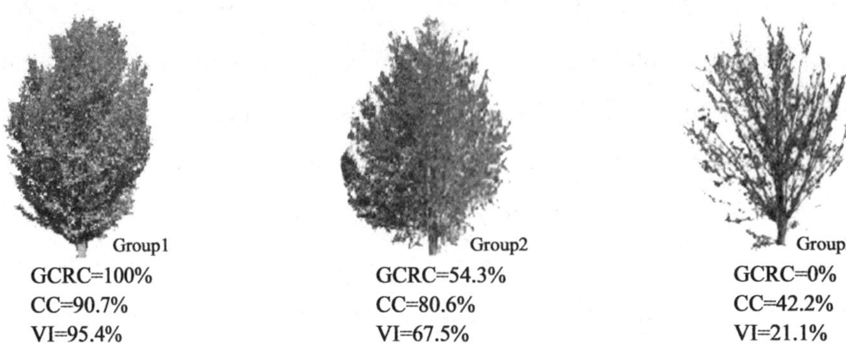

Fig. 10 Model samples of ground-based image for group1, group2 and group3 and related GCRC, CC and VI values.

An outline of the research area (circled area), meteorological data, GCRC index and groups of classification were given in Fig. 1. In the figure, GCRC was scaled into 5 levels, and respectively represented GCRC of 100, 60—90, 40—59, 1—39 and 0. Every sample, with a consistent scale mark, was located on the map in the figure. It can be observed that the classification result was consistent with the research sites. For the first group, the DC was more than 30 km, the second group was from 4 km to 20 km, and the third group was less than 4 km, which was in accordance with the gradient of wind and precipitation. It can be seen that the further the sample tree was from the coastline, the slighter the damage by Typhoon 0613 was according to the indices by ground-based digital image analysis.

From Fig. 1, it is easy to see that there is a number gap between scale 100 and scale 69～90 and it is not difficult to find discontinuous topography between Miyano and Mitani-tokusa which is located in the canyon. This discontinuous topography formed a natural barrier for the trees, protecting them from strong wind blown by typhoon, so that there was almost no sign of damage to ginkgo trees by Typhoon 0613 in this area. If there were no effect of this discontinuous topography, the damaged ginkgo trees might spread far inland and the number gap would not exist.

3.5 Meteorological data and relation analysis

Since the center of Typhoon 0613 brushed the southwest corner of Yamaguchi Prefecture, there was a tendency of wind speed reduction from southwest to northeast in Yamaguchi Prefecture according to the data from AMeDAS. Fig. 11 shows an inverse exponential function relationship between average wind speed and ADC with R^2 equaling 0.723. The investigated area of this research also ran from southwest to northeast and had a tendency of decreasing wind speed from Ube in the southwest with a maximum wind speed of 27 m/s to Tokusa in the northeast with a maximum wind speed of 8 m/s (Table 1) during hit by Typhoon 0613. A similar tendency was expressed for the above-mentioned indices for vigor of ginkgo trees in the investigated area. In brief, we cannot affirm that there is no relationship between the vigor of ginkgo trees and Typhoon 0613 blowing.

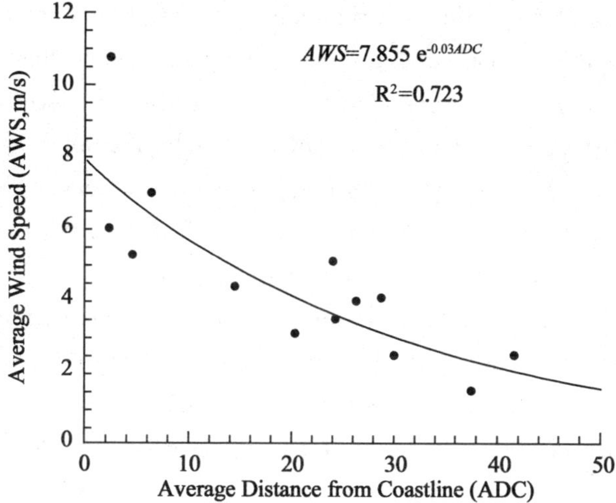

Fig. 11 Relationship between average wind speed (AWS) and average distance from AMeDAS stations to the coastline of west, southwest, south and southeast (ADC) during hit by Typhoon 0613.

According to data from AMeDAS and HPISYP, the precipitation was only 24 mm during Typhoon 0613's hit in Yamaguchi, and only 8.5 mm from Sep. 18th 2006 to Oct. 31st 2006, which is the minimum record of 44 days after hit by strong typhoon (≥15 m/s) for past 40 years (1967—2006). On the other hand, the precipitation during and after Typhoon 0418's hit was so great that it induced regional flooding (Table 2). Meanwhile, almost no such kind of damage to the vigor of ginkgo trees occurred in Yamaguchi after the hit of Typhoon 0418. Evidently, it can be thought that heavy rainfall can counteract damage by typhoons like Typhoon 0613. Ginkgo trees' vigor is damaged by strong wind and lower precipitation together in Yamaguchi City during Typhoon 0613's hit. In other words, it is mainly caused by water stress induced by strong wind from Typhoon 0613.

Table 2 Related precipitation data for Yamaguchi during Typhoon 0613 and 0418

	During Typhoon 0613	During Typhoon 0418
Monthly precipitation in Sep. (mm)	176.0	401.0
Monthly precipitation in Oct. (mm)	5.5	187.5
Precipitation in the day typhoon hit (mm)	24.0(Sep. 17th, 2006)	111.0(Sep. 7th, 2004)
Precipitation for 44 days after typhoon hit (mm)	8.5	440.5

4 Conclusion and discussion

Typhoons are one kind of disaster that can lead to serious damage to forests, trees and shrubs. Besides mechanical damage, vigor reduction is another kind of damage by strong typhoons like Typhoon 0613, which is characterized by discoloration and defoliation of ginkgo crowns accompanying with not-eye-catching branch or twig dieback. By component analysis, leaves on damaged ginkgo trees are composed of leaves with different scorch areas at the time of investigation. Results show that the further they are from the coastline, the fewer scorched leaves, and the closer they are to the coastline, the more scorched and dried leaves. The relationships between DC and indices of GCRC, CC and VI show a similar tendency that the further they are from the coastline, the smaller the damage by Typhoon 0613 is and the bigger the GCRC, CC and VI are. Based on the multi-analysis of this research, similar tendency has been found. Because of the low land productivity in coastal area, the landscape trees should be affected by more complicated factors, in which the salt injury may be one of the serious damage factors (Boyce, 1954; Griffiths and Qrians, 2003; Okinaka and Sugahara. , 1990; Munns, 1993).

Since the 1980s, a gradually increasing number of researches on image analysis have been carried out (Wang, 2006). Most of them were focused on the areas of space-borne or air-borne image analysis for vegetation. Ground-based digital image analysis was also mainly limited to the canopy of plants (Bréda, 2003). Almost no research has been found on typhoon damaged tree crown studies by sideward image, which cannot be detected by space-borne or air-borne equipments. Although there were some reports on typhoon damage with sideward photo as it is, they are not really quantitative researches. In this paper, ground-based digital image analysis was applied in the quantitative research of ginkgo trees' vigor damaged by Typhoon 0613, and was characterized by analyzing sideward image of entire crown. Compared with the common sampling method, it is more effective in application to field measurement of the vigor of trees damaged by severe typhoon with less labor and less time requirement. It may be an alternative tool to be used in estimating or evaluating the degree of damage by typhoons like Typhoon 0613.

As early as in 1805, Salisbury noted that great leaf injury occurred when rain was not associated with strong wind. A universal lower rainfall, averaging 21.7 mm in the investigated area during the hit by Typhoon 0613, reveals that the ginkgo tree's vigor reduced by Typhoon 0613 is just like Salisbury's note. That is why almost no such damage was found in Yamaguchi during Typhoon 0418, which was characterized by heavy rainfall accompanied by a high wind speed even greater than that of Typhoon 0613. It can be said that the damage to the crown of ginkgo trees by Typhoon 0613 is caused by water stress induced by strong wind.

It is observed that lower vigor status of ginkgo trees, especially at limited site condition, seems not caused by one hit. It is more common that before they perfectly recovered from one hit by a storm another hit occurred. This kind of continued damages induced the asymmetric crown of ginkgo trees in the sites with limited conditions and should affect the vigor status to respond the further serious hit by storms like Typhoon 0613.

5 Acknowledgements

We would like to express our gratitude to the Research Lab. of Environmental Ecology, Faculty of Agriculture of Yamaguchi University for providing research conditions and thank all members who help us in our field studies.

References

[1] Boyce S. G. The salty spray community[J]. Ecological Monograph, 1954, 24(1): 29-67.

[2] Bréda N. J. J. Ground-based measurements of leaf area index: A review of methods, instruments and current controversies[J]. Exper. Bot., 2003, 54(392): 2403-2417.

[3] Cullen S. Trees and wind: Wind scale and speeds[J]. Arboriculture, 2002, 28(5): 237-242.

[4] Doswell C. A., Burgess D. W. On some issues of United States tornado climatology[J]. Monthly Weather Review, 1988, 116: 495-501.

[5] Griffiths M. E., Orians C. M. Reponses of common and success ional heath land species to manipulated salt spray and water availability[J]. Amer. J. Bot, 2003, 90(12): 1720-1728.

[6] Guan Wenbin, Li Chunping, Li Shifeng, *et al*. Improvement and application of digitalized measure on shelterbelt porosity[J]. Cn. J. Appl. Eco. 2002, 13(6): 651-657. (in Chinese)

[7] Hayes Ed. Patterns of tree failure[J]. Tree Care Indus., 1999, 10(4): 37-42.

[8] Hennessey J. P. A critique of "trees as a local climatic indicator"[J]. J. Appl. Meteorol., 1980, 19: 1020-1023.

[9] Iwaya K. Studies on Growth Diagnosis of the Rice Plant Canopy by the Optical Measuring Method[D]. The United Graduate School of Agriculture Science of Tottori University, 2003.

[10] Kenny W. A. A method for estimating windbreak porosity using digitalized photographic silhouettes[J]. Agric. For. Meteorol., 1987, 39: 91-94.

[11] Wade J. E., Wendell Hewson E. Trees as a local climatic wind indicator[J]. Appl. Meteoro., 1979, 18: 1182-1187.

[12] Laliberte A. S., Rango A., Fredrickson, *et al*. Separating Green and Senescent Vegetation in Very High Resolution Photography Using Intensity-hue-saturation Transformation and Object Based Classification[C]. 2006 Annual Conference of American Society for Photogrammetry and Remote Sensing, 2006.

[13] Munns R. Physiological processes limiting plant growth in saline soils: Some dogmas and hypotheses[J]. Plant, Cell and Environ., 1993, 16: 15-24.

[14] Purcell L. C. Soybean canopy coverage and light interception measurement using digital imagery[J]. Crop Sci, 2000, 40: 834-837.

[15] Okinaka T., Sugahara M. Wind tunnel experiments on the effect of wind blow and adhering salt in salty wind damage on landscape trees[J]. Tech. Bull. Fac. Hort. Chiba Univ., 1990, 43: 121-128. (in Japanese)

[16] Redfern D. B., Boswell R. C. Assessment of crown condition in forest trees: Comparison of methods, sources of variation and observer bias[J]. For. Eco. Manage., 2004, 188: 149-160.

[17] Richardson M. D., Karcher D. E., Purcell L. C. Quantifying turfgrass cover using digital image analysis [J]. Crop Sci., 2001, 41: 1884-1888.

[18] Robertson A. The use of trees to study wind[J]. Arboric. J., 1987, 11: 127-143.

[19] Robichaud E., Methven I. R. Tree vigor and height growth in Black Spruce[J]. Trees, 1991, 5: 158-163.

[20] Salisbury R. An account of a storm of salt[J]. Linn. Soc. London Trans., 1805, 8: 286-290.

[21] Shimizu Y. Selection of proper landscape tree for costal region (I)—the region salt wind damage easily occur and the distance from coastline[J]. Quart. Hokaidau Fore. Res. Inst., 2004, 134: 16-20. (in Japanese)

[22] Solberg S. Crown density assessments, control surveys and reproducibility[J]. Environ. Monit. Assess., 1999, 56: 75-86.

[23] Takahashi H., Tani H. Study on the interaction between wind and trees in an Urban Area[J]. J. Agr. Met., 1981, 37(3): 239-243. (in Japanese)

[24] Wang Hsueh-Ching, Lin Teng-Chiu. Decisions affecting estimations of understory light environments during photograph acquisition, storage and analysis using hemispherical photography[J]. Taiwan J. For. Sci., 2006, 21(3): 281-295.

[25] Wan Meng, Pan Cun-de, Wang Mei, *et al*. Application of the digitized measurement on windbreak porosity of farmland shelter-forests[J]. Arid Land Geography, 2005, 28(1): 120-123. (in Chinese)

[26] Yamamoto R. Protection of fruit tree against the wind damages[J]. J. Agr. Met., 1979, 35(3): 177-187. (in Japanese)

[27] Zhu Weihua, Xie Liangsheng. The effect of typhoon on landscape trees and solving method in Shenzhen, China[J]. J. Guangdong Landscape Architecture, 2001(1): 25-28. (in Chinese)

[28] Zierl B. A simulation study to analyze the relations between crown condition and drought in Switzerland[J]. For. Eco. Manage, 2004, 188: 25-38.

(原文发表于 *Journal of Nature Disaster Science*, 2008, 30(1):45-53)

第二部分　气象灾害

Responses of some landscape trees to the drought and high temperature event during 2006 and 2007 in Yamaguchi, Japan

WANG Fei[1], Haruhiko Yamamoto[2], Yasuomi Ibaraki[3]

(1. The United Graduate School of Agriculture Science, Tottori University, 9-5-101, Akatsuma, Yamaguchi, Japan. C5417@yamaguchi-u.ac.jp, Tel & Fax: 0081-083-933-5833; 2. Faulty of Agriculture, Yamaguchi University, 1677-1, Yosida, Yamaguchi, Japan; 3. Faulty of Agriculture, Yamaguchi University, 1677-1, Yosida, Yamaguchi, Japan)

1 Introduction

In the surroundings of atmospheric CO_2 increasing and the persistent rising of air temperature in the past 20 years, the annual mean temperature in Japan got higher like the other areas in the world. The significant characteristics of climate change manifested that the numbers of abnormal lower air temperature decreased and extreme higher air temperature (>35 ℃) strikingly increased recent years in Japan (Kurihara, 2007). Although it was difficult to find significant difference of precipitation from normal (Kurihara, 2007), it indicated a trend of raising the number of days of no-rain (Kimoto *et al.*, 2005) and the days of heavy rain over 100 mm or 200 mm (Kurihara, 2007). Therefore, a tendency of adding probability of the extreme weather events characterized by higher temperature and both lower precipitation and higher precipitation would be expected (Meehl and Tebaldi, 2004). Under the condition of no increasing of the total global precipitation, the increasing of rainfall in one region implies the reduction of the precipitation in other areas. In the same region, the seriously positive gain of rainfall in a period of time seems to induce the coming of dry period. The disproportional changes in the upper end of the precipitation frequency distribution in United States, most area of Canada and northeast Mexico may be one special case (Groisman *et al.*, 2007). The drought in 1994 after the year with heavy precipitation in 1993, the persistent less rainfall after the Typhoon 0613's hit and the extreme weather event in 2007 should be another in Yamaguchi, Japan.

In Japan, the perfect irrigation system increased the ability to prevent the paddy rice fields from catastrophic drought so that the rice-crop index in most prefectures of Japan remained near 100% to normal during the extreme weather event in 2007. But most of forest and landscape trees were still in the condition of rainfed growing and showed different abnormalities in the same period. Especially, the extreme weather event of high temperature and low precipitation might affect the plants around invisibly, perhaps injure them (Ciais *et al.*, 2005), even the perennial trees (Pichler and Oberhuber, 2007). For example, in 1994 the paddy rice in west Japan (Yamamoto *et al.*, 1996) and forest trees (Kotani, 1997) suffered from the extreme droughty and hot weather. Historically, in the early 1930's severe drought at the central states of United States and the unusually dry weather in Australia in 1965, many trees appeared early defoliation, leaf scorching, discoloration and so on(Kozlowski, 1976). These kinds of extreme climate even trigger or accelerate the tree mortality (Guar and Taylor, 2005), forest decline (Jurskis, 2005), forest defoliation (Zierl, 2004), reduction of radial growth of trees (Pichler and Oberhuber, 2007), wide-spread primary productivity reduction (Ciais *et al.*, 2005; Barber *et al.*, 2000) particularly in Mediterranean region (Bussotti and Ferretti, 1998), Australia (Jurskis, 2005) as well as the specific area and limited sites (Van der Werf *et al.*, 2007) in some countries.

Plants respond to environmental stresses from unfavorable extreme conditions with characteristics of species-specific and in different patterns. For example, plants in arid habitats are normally exposed to a range of interacting stress and tend to develop deep and/or extensive root system and high value of root/shoot ratio (Fitter and Hay, 2002; Kramer, 1983). As the response to unfavorable extreme weather conditions such as water stress and high temperature etc., plants often appear chlorosis and necrosis (Treshow, 1970; Durbin, 1978). Many tree species also tend to reduce their leaf surface area (Orshan, 1954) by shedding leaves and receive less radiation energy by changing their leaf colors during extreme dry and hot summer days (Kozlowski, 1976; Fitter and Hay, 2002).

Following the abnormal droughty spring, hot and dry summer in 2007 in Yamaguchi, many landscape trees showed abnormal vigor status, especially the trees planted on coarse sand soil, rock mountain site and the site with root growing limitation etc. Many Kousa dogwood trees appeared leaf necrosis on tip and margin also from late August so that the crown of them were made discolored in different scales. Leaves on particularly the leader and upper crown branches of sweet gum trees began to turn to dull colored from mid-September and reddish-brown or purple by mid-October. Some sasanqua trees also fell all of their leaves in late August. During the flower season, the number of flowers on these trees were significantly less than that in 2006 season. A lot of Japanese red pines (*Pinus thunbergii L.*) on mountain sites died, whose needles turned brown first. Some deciduous tree species dropped partial leaves early from late August. The other tree species may be affected by it without visible signs.

The phenomena were measured and described by photo-image pixel analysis supplemented with visual numbers counting and spectral reflectance measuring. As a repaid and non-destructive approach, image pixel color system analysis (Adamsen *et al.*, 1999; Kawashima and Nakatani, 1998; Iwaya and Yamamoto, 2005; Suzuki *et al.*, 1995; Cai *et al.*, 2006) and spectral reflectance analysis (Carter *et al.*, 2001; Thorhaug *et al.*, 2006) used in measuring plant characters can be found in many reports. Image color analysis has also been used in estimating the temporal trends of flower numbers (Adamsen *et al.*, 2000). In this study, image pixel analysis as used in determining leaf necrosis of dogwood, partial crown discolor of sweet gum tree as well as flowering status of sasanqua triggered by extreme weather events in 2007 and Typhoon 0613. The spectral reflectance characteristics of sweet gum and dogwood leaves were analyzed by using spectral measuring method. All of them were aimed at showing the facts of shock to the landscape trees affected by extreme weather event.

2 Materials and methods

2.1 The meteorological data

The annual, monthly, daily meteorological data from 1967 to 2007 for Yamaguchi and other 150 observatories were obtained from Automated Meteorological Data Acquisition System (AMeDAS) of Japan, and the first ten records of maximum or minimum value during these years. The investigated trees were around an area 0.43 km to 1.26 km from the Yamaguchi meteorological observatory. The data were analyzed and described by common statistic parameters.

2.2 LNAP and $NDVI_{765/679}$ of dogwood leaves

The sampled dogwood trees were about 6 years old, 3 to 4 meters high, planted at the site on an ancient riverbed (Sakaue and Ijiri, 1972; Miura and Ono, 1972) area and the former athletic track around a park in Yamaguchi City. The number of sampled trees totaled 61 individuals. In the study, the image G/R (Green/Red) value and LNAP for leaves were determined by analysis of scanned images and 560 leaves were scanned with a flat bed scanner (Canon d125u2).

LNAP was measured by getting pixels of overall leaf and the green part for each leaf. The LNAP was calculated by Formula (1).

$$\mathrm{LNAP}=100-\frac{\mathrm{PGAL}}{\mathrm{POL}}\times 100 \tag{1}$$

in which, PGAL is the pixel numbers of green area of the leaf and POL is the pixel numbers of overall leaf.

The spectral reflectance for leaves with varied LNAP was measured by a radiometer EKO-MS720. Its resolution of spectra is 10 nm, interval of wavelength 3.3 nm and the specified wavelength ranges from 350 — 1050 nm. Leaves were measured with special method at indoor environment and the radiometer was mounted on a tripod 30 cm above the sample leaves. The sample leaves were smoothly filled in a tray in 20 cm×30 cm×4 cm size and vertically measured under 40 W incandescent lamp light. 25° of Field of View (FOV) was selected and the area coverage was 139 cm^2. Measurements were controlled by a piece of white paper corrected by standard white board of barium chloride and 2 duplications for each sample were conducted at different position of the tray. The spectral reflectance was used to calculate the $NDVI_{765/679}$ by Formula (2).

$$NDVI_{765/679}=\frac{NIR_{765}-VIB_{679}}{NIR_{765}+VIB_{679}} \tag{2}$$

where, NIR_{765} is the spectral reflectance at 765 nm in near-infrared region, and VIB_{679} is the spectral reflectance at 679 nm in visible region.

2.3 Analyzing discolored crowns of sweet gum trees

During the abnormal meteorological event, a lot of investigated sweet gum trees discolored from top to base of crown. The sweet gum trees analyzed in the study were single landscape trees along the high way or street. They were visually scaled into classes of entire crown green, less than 1/2 crown discolored and more than 1/2 crown discolored. The number of every scale was visually counted in situ.

In the study, the G/R value in RGB system was used to express the discoloration of the sweet gum crown. The images for calculation of G/R_{cs} (G/R value of crown sections) were taken outdoors by using a CCD digital camera (Canon ixy60) with 5.03 million pixels after discolor symptoms perfectly appeared. The crown images were equally divided into 10 sections and the G/R_{cs} values were determined by Formula (3), which was a proportion of green value of crown section to red value of the same crown section.

$$G/R_{cs}=\frac{\sum_{i=0}^{255}N_i\times i/\sum_{i=0}^{255}N_i}{\sum_{j=0}^{255}N_j\times j/\sum_{j=0}^{255}N_j} \tag{3}$$

where, G/R_{cs} is the G/R value of crown section, N_i is the pixel number in i (green) gradation, $i=0,1,2,\ldots,255$ and N_j is the pixel number in j (red) gradation, $j=0,1,2,\ldots,255$.

The G/R_{vcs} is one kind of G/R_{cs} value for crown sections divided from base to top of the crown. The G/R_{hcs} is also a type of G/R_{cs} value for crown sections divided from left or leeward to right or windward of the crown. Before getting pixel data, the image was hand prepared by eraser of Photoshop to remove the objects except the objective crown. Then the Red (R), Green (G) values were read from the average histogram value of individual crown extracted by tools of Photoshop.

It is observed that the Relative G/R_{cs} (RGR) of injured crowns decreased from proximal to distal in the manner of logistic differential function as Formula (4).

$$\mathrm{RGR}(n) = \frac{k}{1 + \mathrm{e}^{a-m}} \tag{4}$$

where, RGR(n) stands for relative G/R_{cs} [formula (5)] index at n section, k is the maximum value the G/R_{cs} can reach, the r and a are constant, and n is the number of crown sections.

$$\mathrm{RGR}_i = \frac{100 \times (G/R_i - G/R_{\min})}{G/R_{\max} - G/R_{\min}} \tag{5}$$

where, G/R_i is the G/R value for i section, G/R_{min} and G/R_{max} are the minimum and maximum G/R values in all sections respectively.

2.4 Estimation of flowering pixels and flower number of sasanqua

The showy red/pink flowers of sasanqua trees become proper characters to analyze their response to the extreme weather events. In thc study, the photo images were taken from the same sasanqua tree on the same day in 2006 and 2007 flower season respectively. The flower number was visually counted from image and the flower pixels were directly read from histogram value of Photoshop in which flowers were extracted by selection tools.

3 Results and analysis

3.1 Characteristics of climate change in Yamaguchi

From the beginning of the AMeDAS observation from 1967 in Yamaguchi, the annual mean temperature drew a fluctuated increasing line (Fig. 1a), with a linear function ($R^2=0.597$), like the most of other cites in Japan. By average, the air temperature increased 1.68 degree centigrade from 1967 to 2007. Although, the annual precipitation almost remained at the same level (Fig. 1a), it turned to raise the fluctuations of their standard deviation in recent 21 years especially in the second half of a year (Fig. 1b). The tendency of amplification of deviation for precipitation even appeared in more than

60% of 150 central observatories in Japan. During the global climate change, not only the heavy rainfall has increased in some regions (Kurihara, 2007; Matsumoto and Yamamoto, 2007), but also the probability of extreme weather events have raised such as severe drought (Neilson and Drapek, 1998). Integrating the weather events from 1967 to 2007, there appeared a tendency of more negative extreme spread value of the precipitation-temperature proportion (P/T), less positive extreme of it in Yamaguchi from 1987 to 2007 (Fig. 1c). It was indicated that a trend of more years with high temperature and low precipitation and less years with low temperature and high precipitation occurred. This kind of trend even appeared in entire Japan. Meanwhile, the occurrence of strong typhoons, gust wind speed over 33 m/s, increased during 1987 to 2007 compared with that from 1967 to 1986 and the standard deviation of them also raised (Fig. 1d line). Both the mean value of precipitation 44 days after strong typhoon's hit and rainfalls in the day strong typhoon's hit were lower during 1987 to 2007 than 1967 to 1986. The standard deviation for both of them got larger during 1987 to 2007 than from 1967 to 1986 (Fig. 1d histogram). Therefore, it seems to increase the probability of extreme weather event of high temperature and low precipitation in local area under this situation (Bachelet *et al.*, 2001).

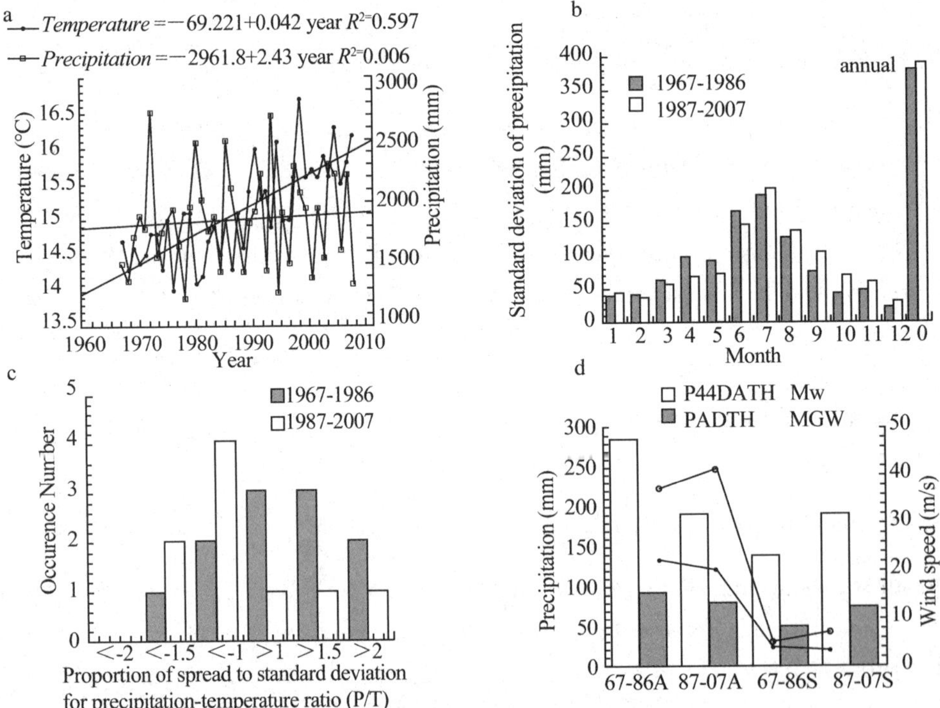

Fig. 1 Characteristics of climate change in Yamaguchi, Japan: 1a represents the temporal series of annual mean temperature and precipitation from 1967 to 2007, which appeared a slant and a level line for temperature and precipitation respectively. 1b shows the yearly and monthly distribution characteristics of standard devia-

tion of precipitation in 1967—1986 and 1987—2007 for Yamaguchi City and indicates a characteristic of larger precipitation standard deviation at all months in the second half year. 1c presents the occurrence number in different folds to the standard deviation value of precipitation-temperature ratio for Yamaguchi. It occurred more times in 1987～2007 than in 1967～1986 in the area of minors value of spread and standard deviation ratio and less times in the area of positive value of spread and standard deviation ratio. 1d is a graph of max gust wind speed (MGW) and max wind speed (Mw) for strong typhoons whose gust wind speed is over 33 m/s, the precipitation in the day typhoon's hit (PADTH) and the rainfall during 44 days after typhoon's hit (P44DATH).

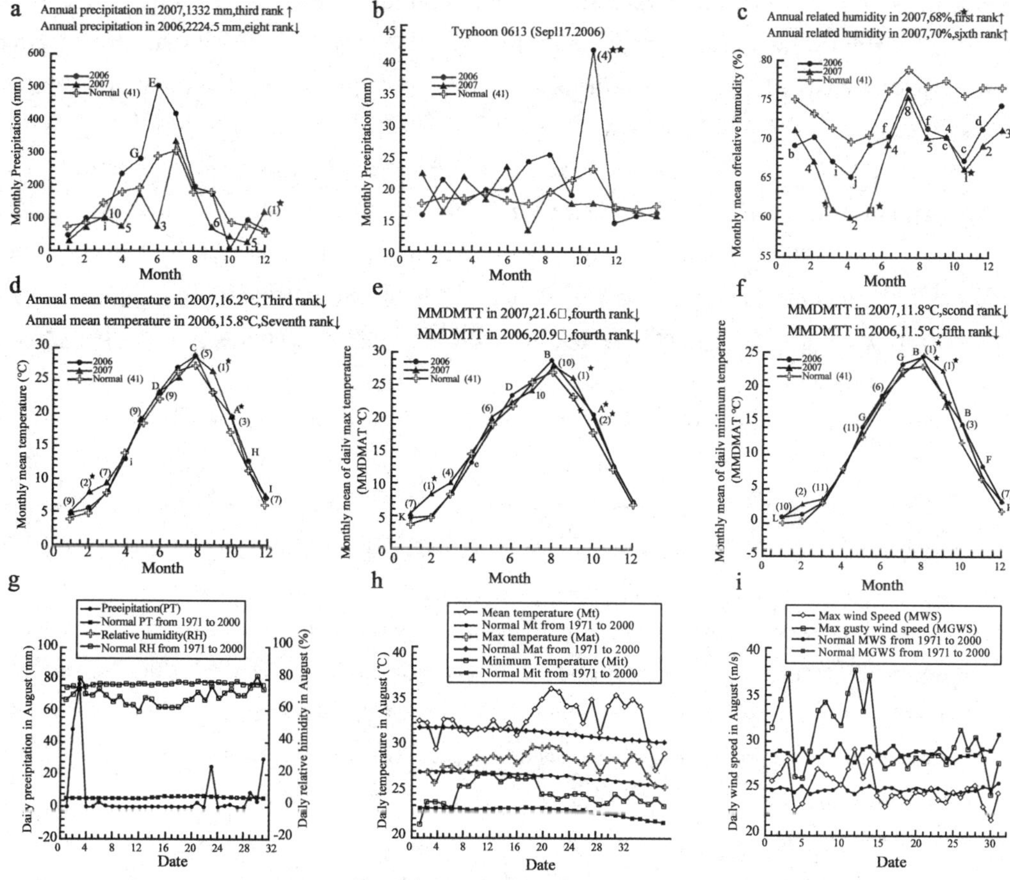

Fig. 2 Extreme weather events during 2006 and 2007 and their comparison to normal years during 1967—2007 from 2a to 2f, in which, the extreme record in 2006 is marked with capital letters for descending order, small letters for ascending order; and numbers for ascending order, numbers with bracket for descending order in 2007. The value with * mark means its ratio of spread value to standard deviation is more than 2.0. The characteristics of August in 2007 are described from 2g to 2i. They present daily precipitation, daily relative humidity, normal values (1971—2000) of precipitation and relative humidity in 2g; mean temperature, max temperature, minimum temperature and normal values of them in 2h; max wind speed and gusty wind speed in August and normal values of them in 2i.

3.2 Extreme weather events during 2006 and 2007

In the year 2007, the abnormal weather with uneven rainfall occurred in many areas over the world. Severe drought happened in southeast United States, southeast Australia and northeast China, while the extreme rainy weather took place in Europe (Meteorological Agency of Japan, 2007a, 2007b, 2007c, 2007d, 2007e, 2007f). The weather characteristic during the year 2006 and 2007 in Yamaguchi City, Japan also showed a sharp contrast (Fig. 2). The plentiful precipitation in East Asian rainy season (Fig. 2a), strong Typhoon with less rainfall in mid-September, and higher temperature associated with almost no rainfall in October became the extreme weather events in 2006 (Fig. 2a, 2b, 2d, 2e). The less rainfall after strong Typhoon 0613's hit almost persisted through the entire 2007 growth season (Fig. 2a). The extreme weather event in 2007 was expressed for its less and uneven precipitation, lower relative humidity and higher temperature than that of normal years (1967 to 2007, Figs. 2a, 2c, 2d, 2e, 2f). It took the first rank of annual minimum relative humidity (Fig. 2c), third rank of minimum annual precipitation (Fig. 2a) and third rank of annual maximum value of mean temperature during these 41 years (Fig. 2e). Particularly, it made new records of many monthly meteorological variables (Fig. 2).

The extreme weather from August seventh to eighteenth can be considered as the key of extreme weather event in August 2007 in Yamaguchi. Although hit by Typhoon 0705 on August 3, 2007, its max wind speed was lower than 10 m/s (Fig. 2i) accompanying with heavy daily precipitation (77.5 mm, Fig. 2g) and not very higher temperature (daily highest temperature 29.8 ℃, Fig. 2h) in Yamaguchi. Then, more than 10 days' anticyclone weather occurred. During these days, as the average temperature and minimum temperature maintained higher than the normal years (1971—2000), the max wind speed and gusty wind speed reached the peaks at 9.1 m/s and 17.7 m/s (Fig. 2i) associated with no rainfall (Fig. 2g). Meanwhile, the relative humidity drew a "U"-shaped curve (Fig. 2g) and the period of no rain persisted seventeen days. All of these led to a foehn like weather.

Although no unique definition of the extreme weather event has been found in the related fields, it is an event with small probability to be confirmed. Here the extreme weather events were considered as the events that the spread value (SPV) of meteorological variables reached 2.0 times more (or less) than that of standard deviation (STDEV). From the end of the strong Typhoon 0613's hit to the end of 2007, the ratio of SPV/STDEV for fourteen meteorological variables exceeded 2.0 (with * mark in Fig. 2) and most of them made a new record. It suggested that the meteorological environment in 2007 in Yamaguchi, Japan was characterized by dry in almost all of the first eleven months and hot in February, August, September and October. The annual precipitation was lower and uneven, which was 71.6% of normal year and only 60.1% for the first nine months. The related year data for entire Japan is 87.6%, with the lowest one 55% or so (Table 1). In Yamaguchi, the annual mean temperature accounted for 107.3% of the normal year and 106.7% in entire Japan. The highest one reached 113.5% of normal year and only one observatory was lower than normal year (2007/normal<100%) in 2007 in entire Japan. Although, the precipitation reached a height of making new record of maximum monthly precipitation in Dec. in Yamaguchi, it was too late to prevent the trees from responding to the hot, dry environment so as to various symptoms appeared on many landscape trees, especially the trees planted on the poor soils and the site with root growing limitation etc.

Table 1 Spread and ratio of annual precipitation and temperature in 2007 to normal year

	Precipitation (mm)		Temperature (℃)	
	Yamaguchi	Japan	Yamaguchi	Japan
Ratio (%)	71.6 (60.1 *)	87.6	107.3	106.7
Max ratio (%)	—	137.4	—	113.5
Min ratio (%)	—	55	—	98.9
Spread	517.1	215.9	1.1	0.75

Note: The data of Japan are mean values from 150 observatories. The data with * meant data from the first nine months.

3.3 Responses from some landscape trees

By comparison, the responses from landscape trees to extreme climate events are more visible than the meteorological variables. But they are not easy to be measured as the latter. Although the different tree species may respond water stress and hot, dry wind injury differently, the responses always make them varied toward decreasing water loss and accepting less radiation energy. Particularly in summer, root growth is often reduced and stopped by soil water deficit (Kramer, 1983). During the hot and dry summer trees can appear premature discolor on leaves on particularly the leader and upper crown (Leaphart, 1959). Similar symptoms were noted in 2007 and leaves of sweet gum trees began to turn leaf color from dull in mid-September to reddish-brown or purple by mid-October (Fig. 3b). In contrast, the normal growth (Fig. 3a, non discolored leaves at left half of crown) of sweet gum trees before mid-September benefited from the plentiful precipitation in the first eight months in 2006 (Fig. 2a). Although no such kind of dry and hot summer arose in 2006 as in 2007, the strong Typhoon 0613 led many of them asymmetrically discolored from windward to the leeward (Fig. 3a) in mid-September. Described by image pixel analysis, the horizontal differential equation of image RGR value for the crowns of sweet gum trees in 2006 showed an inverse logistic equation from leeward of the crown to windward and an approximate horizontal line for the vertical differential (Fig. 3e). On the opposite, it almost presented an approximate level line for horizontal differential equation for 2007 and an adverse logistic equation by vertical differential from the bottom of crown to the top (Fig. 3f).

In nature, many plants adapted the droughty environment by increasing their root system and developing high value of root/shoot ratio (Kramer, 1983). High values of this ratio also can be achieved by reduced transpiring surface or leaf area (Fitter and Hay, 2002; Kozlowski, 1976). The investigation from 381 stocks of sweet gum trees, planted in Yamaguchi City, indicated big difference between pruned and non-pruned trees, even the trees planted at different sides of the same road. According to Fig. 3d, seldom of pruned sweet gum trees (Fig. 3c) became partial discolored or entire crown discolored with the percentages of 90.1%, 7.1% and 2.8% respectively for crowns of entire green, <1/2 discolored and >1/2 discolored. Most of the partial discolored and entire discolored trees were non-pruned (Fig. 3b) trees with the percentages of 22.8%, 44.8% and 32.4% respectively for the same scales (Fig. 3d). It seemed that the tree pruning increased the ability to prevent themselves from injury by extreme weather event by changing their root/shoot ratio. Pruning is a common practice in landscape tree management and it directly results in increasing the root/shoot ratio of pruned trees. For the clear pruning, only the main stem remains, and the root/shoot ratio almost becomes ∞ thereotically. It was not strange that the new pruned sweet gum tree endured more serious hot and dry environment since it directly improved the metablic balance and hormone harmoniousness, especially changed water relations (Kramer, 1983). Based on the result of spectral reflectance measuring, the red sweet gum leaf reflected more energy at red spectral region than that of green leaf while the purple sweet gum leaf reflected more near infrared energy than the green leaf. Only at the visible green region the reflectance from green leaf was greater than that of purple and red leaves (Fig. 3g). It should be less doubt that the premature discolor of sweet gum trees could improve the energy balance.

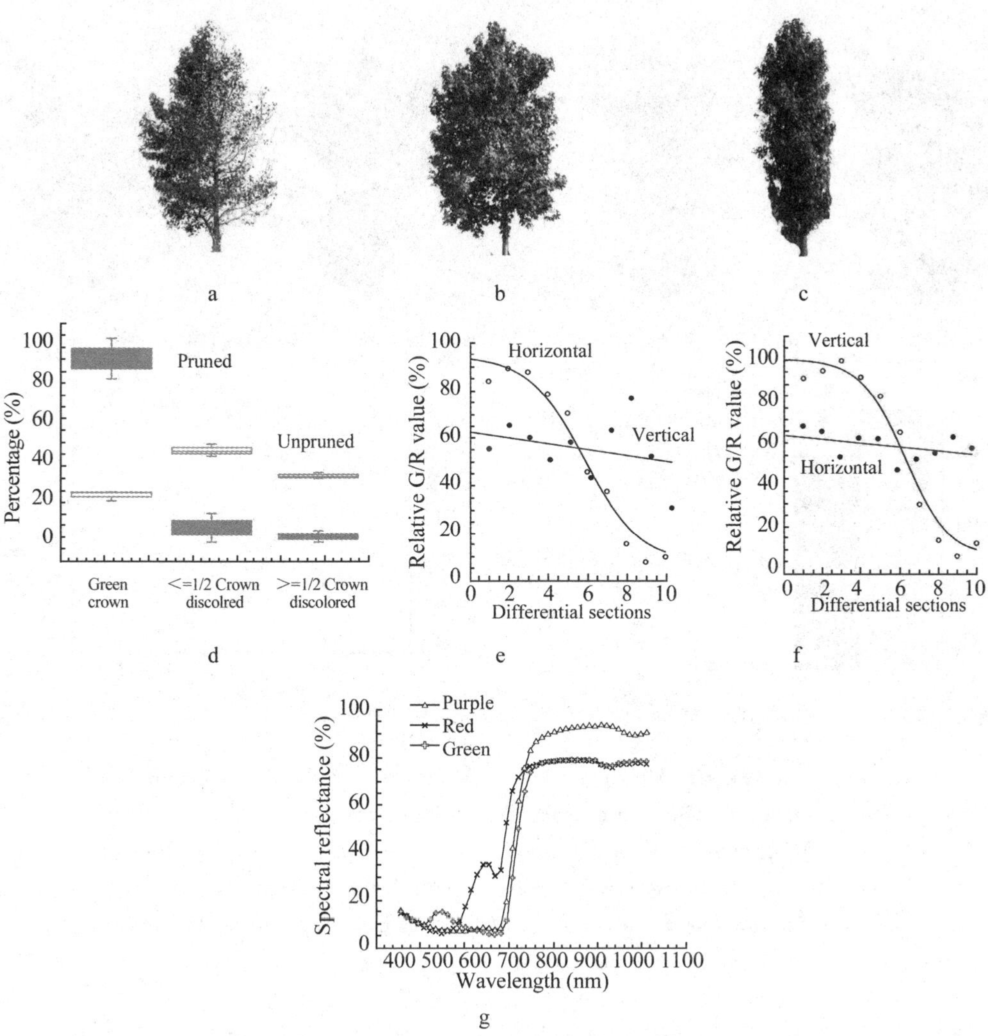

Fig. 3 The abnormal status of sweet gum tree during different extreme weather events in 2006 and 2007, in which the typical crown status of sweet gum tree in 2006 and 2007 was respectively presented in 3a and 3b. Their differential functions of relative G/R (RGR) value from vertical and horizontal orientation were respectively showed in 3e and 3f. The typical crown status of pruned and non-pruned sweet gum trees was respectively presented in 3c and 3b. Their statistical percentages were presented in 3d, where "green crown" is for the entire crown which remains green color, "<1/2 discolored" stands for the sweet gum trees whose discolored part of crown is less than 1/2 and ">1/2 discolored" means the discolored part is more than 1/2. 3g presented the spectral reflectance curves for green (○—○), purple (△—△) and red (×—×) of sweet gum leaves at visible and near infrared region.

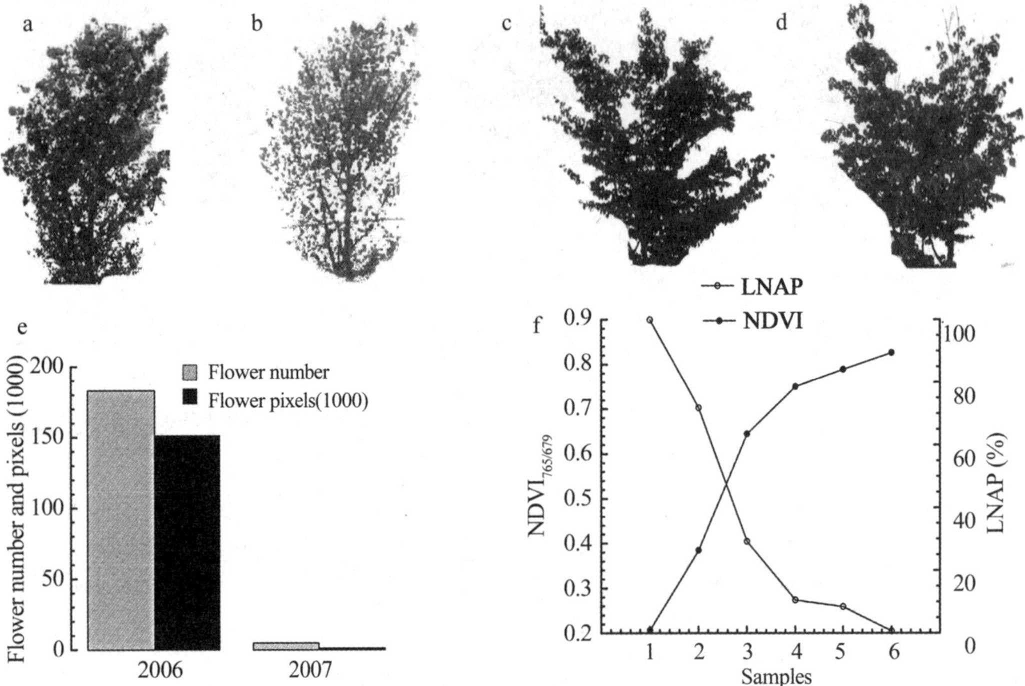

Fig. 4 Different flowering status for the same sasanqua tree in 2006 and 2007 flower season were separately presented in 4a and 4b, and the flower number and flower pixels in 4e. The dogwood almost with no necrotic leaves like the crowns before hit by Typhoon 0613 in 2006 was showed in 4c. In 4d the crown with several leaf necrosis after extreme weather event in 2007 appeared. 4f was the LNAP (○-○) and $NDVI_{765/679}$ (●-●) values of dogwood leaves measured in Oct. 2007, which presented a striking adverse relationship

The most obvious response to the extreme weather event in 2007 was defoliation in late August and fewer flowers in flower season for some sasanqua trees, especially the trees on limited sites. During the 2006 flower season, full blooming flowers (Fig. 4a) for the sampled sasanqua tree might profit from the proper precipitation in summer (Fig. 2a) and normal growth in growing season. After being affected by the extreme weather event, the flower numbers of the sasanqua tree in 2007 season were very fewer (Fig. 4b) for the reason of the summer defoliation. The image flower pixels measured from the same tree in 2006 and 2007 seasons also appeared big difference (Fig. 4e). In 2007 season, no flower existed on upper half of the crown and only fewer flower bloomed at lower half of the crown at investigated time. It suggested that the shading from upper part of the crown benefited the flowers at lower part.

The symptoms of tip and margin leaf necrosis appeared on many dogwood trees on poor site conditions in Yamaguchi City from late August, 2007, which made the crown of them discolored in different scale. The LNAP of sampled trees ranged from near 0

(Fig. 4c) to 100% and indicated that serious leaf necrosis occurred on the leaves of dogwood (Fig. 4d). It presented 41.6% average LNAP by the image pixel measurement of necrotic leaves. About 40% leaf area of injured dogwood tree was reduced through the partial leaf necrosis by the end of August 2007. Even if it was coincidence that the precipitation during the first nine months was about 40% less than that of normal years (Fig. 2a), the relevance between the leaf necrosis of dogwood trees and less precipitation should be less doubt. It indirectly decreased the water (precipitation) requirement and received less radiant energy for the living parts of entire tree. On the base of continuative observation, the dogwood trees that showed symptoms of leaf necrosis in the middle of August in 2008 were less than 1/2 in 2007. The total leaf area reduction in 2008, only 13.2% of the entire leaves, was less than 1/3 of that in 2007.

From Fig. 4f, a significant adverse relationship between LNAP and $NDVI_{765/679}$ value of necrotic dogwood leaves can be seen. It also indicated that more spectral at red region was reflected by severe necrotic leaves than that of normal leaves and received less energy for the living part of the necrotic trees comparad with the normal trees.

4 Discussion

During the extreme climate event in 2007, 20% of the 150 observatories in Japan showed the precipitation less than 80% of normal and the temperature more than 105% of normal. In Yamaguchi, because of the extreme hot and dry weather event in 2006 and 2007, many meteorological variables occurred once in more than 40 years and the complex of these variables may be as long as 50—60 years or even more in Yamaguchi. Although whether it was related to the global warming still remained unclear, it happened in Yamaguchi under the background of climate change. These changes, especially the higher temperature associated with local drought, will endanger the sensitive landscape trees. If the large scale warming still continued, the similar extreme event may be expected much frequently.

Many kinds of meteorological variables affect plants, directly or indirectly, singly and complexly, rapidly and slowly, as well as persistently and temporarily (Maki *et al.*, 1991). Plants will develop normally as long as each of many environmental factors remains within a critical range. Extreme environmental factors affect the plants comprehensively, and the injuries tend to be most serious and the threshold double decreases when more factors are in the stress status. The most evident responses to the drought and high temperature event during 2007 in Yamaguchi appearing on some landscape trees became one of the cases, and only the trees planted on limited sites significantly responded to the extreme weather event.

Leaf scorch emerges on many different kinds of plants but is particularly common on shade-love species and leaves developed under cool conditions and suddenly exposed to the intense heat energy of the summer sun. Leaves are also predisposed and necrose most severe when a period of cloudy, rainy weather promoting rapid top growth is followed by hot sunny days (Treshow, 1970) and there is no enough water supply. During the extreme weather event in 2007 in Yamaguchi, although the precipitation in July got over the normal year, this kind of temporary restore of water supply might make the necrosis of the dogwood leaves more serious in late August for the reason of serious water imbalance. This kind of extreme weather event can be described as the momentary Mediterranean-climate-like event.

The transpiration cooler fail during the serious drought stress seems lethal to plants. The reduction of transpiring surface by shedding of leaves has been considered to be the most important factor in survival of many desert plants (Orshan, 1954). Although many tropical trees are classified as summer deciduous, some of them tend to shed their leaves during dry period(Kozlowski, 1976). They are considered as species with characteristics of environment specific leaf shedding. It was in the late August that many sasanqua trees in Yamaguchi evaded the extreme droughty and hot stress by leaf shedding, which usually did not defoliate in the normal years, at the expense of biomass.

During the growth of trees, if leaves do intercept excessive amounts of solar radiation under conditions of water shortage, survival will depend on convection of excessive heat from the leaf surface of reflection on incident radiation before it can be absorbed (Fitter and Hay, 2002). It was clear that the leaves of sweet gum trees became red or purple resulting in reflecting more light energy in red and infrared ranges than that of green leaf and less radiation energy from direct sunshine was accepted by the discolored trees.

During the extreme weather event in 2006 and 2007, two typhoons directly affect the landscape trees in Yamaguchi. One is the Typhoon 0613 characterized by strong wind associated with less rain. After its hit, many crowns of landscape trees became asymmetrically discolored. By contrast, there were almost no visible impact on the landscape trees in Yamaguchi after hit by Typhoon 0705 characterized by moderate wind and proper precipitation. However, the wind during anticyclone weather in the mid-August in 2007 seemed to play a role of increasing the injury power by hot and dry weather events.

With different genetic characters of leaf structure, the landscape trees of dogwood, sweet gum tree and sasanqua responded to the extreme weather event during 2006 and 2007 variedly in Yamaguchi. Although, the strong wind with less rainfall in 2006 did not lead the evergreen sasanqua to appear significant visible disorder, the dry, hot and windy weather event in 2007 caused their leaf shedding and less flower bloomed in flower season. The leaf necrosis induced by draughty and hot weather event in 2007 may be attributed to the light textural leaflet and persistence of the dogwood leaves. The leaf discoloration from upper to the base of the crown for sweet gum tree seemed one of the striking adaptive characteristics during the extreme weather event in 2007.

Landscape trees were usually selected and planted by their characteristics of ornamental values. Partial of them are aesthetically planted and regenerated at their unfavorable site so as to be sensitive to the environmental changes. Nevertheless, under the proper stress condition, the ornamental value of many landscape trees is even increased. The response of these trees to the extreme weather event may be an effective reference of the climate change. They could become usable resources to study the injury or shock by extreme environment changes, especially for the humid area like Yamaguchi with normal year precipitation more than 1,800 mm.

References

[1] Adamsen F. J., Pinter P. J. Jr., Barnes E. M., LaMorte R. L., Wall G. W., Leavitt S. W., Kimball B. A. Measuring wheat senescence with a digital camera[J]. Crop Sci., 1999, 39: 719-724.

[2] Adamsen F. J., Coffelt T. A., Nelson M., Barnes E. M., Rice R. C. Method for using images from a color digital camera to estimate flower number[J]. Crop Sci., 2000, 40: 704-770.

[3] Bachelet D., Neilson R. P., Lenihan J. M., Raymond J. D. Climate change effects on vegetation distribution and carbon budget in the United States[J]. Ecosystems, 2001, 4: 164-185.

[4] Barber V. A., Juday G. P., Finney B. P. Reduced growth of Alaskan white spruce in the twentieth century from temperature-induced drought stress[J]. Nature, 2000, 405: 668-673.

[5] Bussotti F., Ferretti M. Air pollution, forest condition and forest decline in Southern Europe: An overview [J]. Environ. Pollut., 1998, 101: 49-65.

[6] Cai H. C., Cui H., Song W. T., Gao L. H. Reliminary study on photosynthetic pigment content and color feature of cucumber initial blooms[J]. Trans. CSAE, 2006, 22(9): 34-38. (In Chinese)

[7] Ciais Ph, Reichstein M., Viovy N., Granier A., Ogée J., Allard V., Aubinet M., Buchmann N., Bernhofer Chr, Carrara A., Chevallier F., De Noblet N., Friend A. D., Friedlingstein P., Grünwald T., Heinesch B., Keronen P., Knohl A., Krinner G., Loustau D., Manca G., Matteucci G., Miglietta F., Ourcival J. M., Papale D., Pilegaard K., Rambal S., Seufert G., Soussana J. F., Sanz M. J., Schulze E. D., Vesala T., Valentini R. Europe-wide reduction in primary productivity caused by the heat and drought in 2003[J]. Nature, 2005, 437: 529-533.

[8] Carter G. A., Knapp A. K. Leaf optical properties in higher plants: Linking spectra characteristics to stress and chlorophyll concentration[J]. Amer. J. Bot., 2001, 88: 677-684.

[9] Durbin R. D. Abiotic Diseases Induced by Unfavorable Water Relations in "Water Deficits and Plant Growth, vol. 5", (Kozlowski T. T., ed.)[M]. New York: Academic Press, 1978: 101-107.

[10] Fitter A. H., Hay R. K. M. Environmental Physiology of Plants[M]. Academic Press, 2002: 162-169.

[11] Groisman P. Ya, Knight R. W. Prolonged dry episodes over North America: New tendencies emerging during the last 40 years[J]. Advances in Earth Science, 2007, 22: 1191-1207.

[12] Guar A., Taylor A. H. Drought triggered tree mortality in mixed conifer forests in Yosemite National Park, California, USA[J]. Forest Eco. Manage, 2005, 218: 229-244.

[13] Jurskis V. Eucalypt decline in Australia, and a general concept of tree decline and dieback[J]. Forest Eco. Manage, 2005, 215: 1-20.

[14] Iwaya K., Yamamoto H. The diagnosis of optimal harvesting time of rice using digital imageing[J]. J. Agric. Meteorol, 2005, 60: 981-984. (In Japanese)

[15] Kawashima S., Nakatani M. An lgorithm for estimating chlorophyll content in leaves using a video camera [J]. Ann. Bot., 1998, 81: 49-54.

[16] Kimoto M., Yasutomi N., Yokoyama C., Emori S. Projected changes in precipitation characteristics around Japan under the global warming[J]. SOLA, 2005, 1: 85-88.

[17] Kotani E. The stand structure and diameter increment of Hinoki cypress after hit by summer drought caused by less rainfall in 1994[J]. Application Research of Forest, 1997: 25-28. (In Japanese)

[18] Kozlowski T. T. Water Supply and Leaf Shedding in "Water Deficits and Plant Growth, vol. 4", (Kozlowski T. T., ed.)[M]. New York: Academic Press, 1976: 191-222.

[19] Kramer P. J. Water Relation of Plants[M]. Academic Press, 1983: 370-371, 161-164.

[20] Kurihara K. Current characteristics of abnormal weather and climate change[J]. Tenki, 2007, 54(7): 21-25, 44-45. (In Japanese)

[21] Leaphart C. D. Drought damage to western white pine and associated tree species[J]. Plant Dis. Rep., 1959, 3: 7.

[22] Maki T., Suzuki Y., Kamoda F., Hayakawa S., Tomari K. Meteorological Disaster in Agriculture and the Countermeasure[M]. Yokendo Press, 1991: 110-137. (In Japanese)

[23] Matsumoto J., Yamamoto N. The recent characteristic of precipitation over the world[J]. Tenki, 2007, 54 (7): 26-30. (In Japanese)

[24] Meehl G. A., Tebaldi C. More intense, more frequent, and longer lasting heat waves in the 21st century [J]. Science, 2004: 305, 994-997.

[25] Meteorological Agency of Japan Climate Information, Tenki, 2006, 54(2): 72-73. (In Japanese)

[26] Meteorological Agency of Japan Climate Information, Tenki, 2007a, 54(3): 46-47. (In Japanese)

[27] Meteorological Agency of Japan Cliamate Information, Tenki, 2007b, 54(5): 78-79. (In Japanese)

[28] Meteorological Agency of Japan Climate Information, Tenki, 2007c, 54(6): 38-39. (In Japanese)

[29] Meteorological Agency of Japan Climate Information, Tenki, 2007d, 54(4): 48-49. (In Japanese)

[30] Meteorological Agency of Japan Climate Information, Tenki, 2007e, 54(10): 38-39. (In Japanese)

[31] Meteorological Agency of Japan Climate Information, Tenki, 2007f, 54(11): 34-35. (In Japanese)

[32] Miura T., Ono T. Classification of Topography in Fundamental Land Classification Survey, Ogori Volume of Geomophology, Surface Geology and Soil, ed. by Economical Planning Agency of Japan[M], 1972: 14-15.

[33] Neilson R. P., Drapek R. J. Potentially complex biosphere responses to transient global warming[J]. Global Change Biol, 1998, 4: 505-521.

[34] Orshan G. Surface reduction and its significance as a hydroecological factor[J]. J. Ecol., 1954, 42: 442-444.

[35] Pichler P., Oberhuber W. Radial growth response of coniferous forest trees in an inner Alpine environment to heat-wave in 2003, Forest Eco[J]. Manage, 2007, 242: 688-699.

[36] Sakaue Y., Ijiri T. Soil interpretation in "Fundamental Land Classification Survey, Ogori Volume of Geomophology, Surface Geology and Soil, ed. by Economical Planning Agency of Japan, 1972: 32-33.

[37] Suzuki T. Measurement of growth of plug seedlings by image processing in broccoli[J]. Acta Horticulturae, 1995, 399: 333-343.

[38] Thorhaug A., Richardson A. D., Berlyn G. P. Spectra reflectance of Thalassia testudinum (Hydrocharitaceae) seagrass: Low salinity effects[J]. Amer. J. Bot., 2006, 93: 110-117.

[39] Treshow M. Environment and Plant Response[M]. New York: Mcgraw-Hill, 1970: 22-174.

[40] Van der Werf G. W., Sass-Klaassen U. G. W., Mohren G. M. J. The impact of the 2003 summer drought on the intra-annual growth pattern of beech (Fagus sylvatica L.) and oak (Quercus robur L.) on a dry site in the Netherlands[J]. Dendrochronologia, 2007, 25: 103-112.

[41] Yamamoto H., Suzuki Y., Hayakawa S., Hirayama K. Survey on meteorological characteristics of dry summer and paddy rice damage caused by the drought in western part of Japan in 1994[J]. J. Nat. Disaster Sci., 1996, 15: 11-17. (In Japanese)

[42] Zierl B. A simulation study to analyse the relations between crown condition and drought in Switzerland[J]. Forest Eco. Manage., 2004, 188: 25-38.

(原文发表于 *Journal of Forestry Research*, 2009, 20(3): 254-260)

一些景观树对灾害天气事件的非对称响应

王　斐[1]　张继权[2]

(1. 山东省林业科学研究院，山东，济南 250014；2. 东北师范大学城市与环境科学学院，吉林，长春 130024)

1. 引言

对称性是正常生长的乔灌木枝叶和树冠结构发育的显著特征，而且这种特征通常具有形态上的遗传性[1]。然而，不断增加的近期研究事实表明，这种对称性仅表现为波动性的对称[2]，树木只有在稳定不变的环境中才能发育出对称的枝和树冠[3]。在某些环境条件下，众多内外因素可阻止树木的对称性发育。例如，在盛行风的地方，树木可由于风压而导致树冠变形[3,4]。在近海岸地带的树冠变形被认为起因于盐雾危害[5]，而在高山树线地带的变形树被认为是遭受机械摩擦、雪害或冰晶危害所致。也有人认为强烈的风压伴随严重的水分胁迫才是导致一些树木偏冠的主要原因[6]。事实上，树木偏冠变形的原因往往是许多因素的综合作用，这些综合作用极大地降低了树木响应的临界值。从这种意义上讲，持不同观点的研究人员在理解偏冠的原因方面各有其亮点所在。在研究树木变形或偏冠形成的历史中，大量研究聚焦于环境因子对树木的损害，突出表现在追寻伤害的决定性因子[5]，争议和冲突常常集中在几个物理或化学因素上面。但是，自然环境中的树木通常受制于一系列生物或非生物因子，表现出来的症状往往是树木对外界环境的响应，且因遗传基础不同而异。对那些树枝具有特殊的机械特性的树种而言，机械的驯化似乎对其偏冠更重要[3]。对于那些盐敏感型的树种，盐害似乎是非对称枝叶枯死的主要原因[7]。树木个体高大、空间结构复杂的特性常造成自我的遮挡效应。树冠无遮挡的部位或暴露于灾害袭击的部位受害更加严重。在水分胁迫下，树冠外围或顶部裸露在外的树叶比树冠内部或下部的受害要重[8]。这种持续的自我遮挡常导致景观树明显的偏冠。在灾害天气事件袭击后的恢复期间，不同部位的不对称生长或许是偏冠的另外一个成因。在众多环境因子中，直接或间接地明显改变树木响应特征的因素才应该是决定性的因素。

水分是生命的基础，草本植物中水分占鲜重的绝大部分，在木本植物结构中水分也占鲜重的50%以上[9]。水分参与光合作用过程,维持细胞的膨胀状态,且具有调节温度和辅助养分运输等功能。在极端水分胁迫条件下,许多树木能够以牺牲部分器官为代价来拯救自身[10],诸如落叶、落枝、枯叶和枯枝等[8,9,11,16],尽管不同的植物种表现出了不同的可缩性和多样性。地中海类型的夏季干旱和热带风暴是两种特殊的气象现象,它们常常诱发植物或树木的强烈响应。类似的气象灾害事件对森林和树木的严重扰动可发生于世界许多地区。夏季高温和强台风伴随着无雨和持续的干旱可诱发景观乔灌木树种明显的可视反应,如枯叶、枯枝以及器官脱落[15,17,18]。台风 0613 号在 2007 年夏季干旱对山口市的影响尤为突出。强台风 0613 号登陆时,日本山口地区的最大瞬时风速达到了 42.4 m/s,台风袭击期间的降水量仅有 26 mm,且在风速达到最大时几乎没有什么降雨,紧接着就是长达 1 个多月的少雨期。在 2007 年夏季干旱期间,降水量仅为常年的 61.2%。受其影响,周围的景观树有的表现出了明显的可见症状。此类灾害天气事件持续地单向袭击树冠的一侧诱发了枝条的不平衡生长和偏冠。本研究通过对气象数据的分析、图像解析等,阐述了日本山口市台风 0418 号、台风 0613 号和 2007 年夏季干旱期间灾害天气事件的突出特点,并对一些景观树的非对称响应进行了研究,旨在揭示在既无盛行风也没有严重盐雾影响情况下树木发生偏冠的事例。

2. 材料与方法

2.1 气象数据和相关的指数

研究所在地为日本山口市,调查是沿椹野河流域进行的。本研究的一部分基于对山口测侯所的日本全自动气象数据观测系统所获数据的分析,其地埋位置距海岸约 13 km。从中选取逐日气温、降水量和超过 33 m/s 的最大瞬间风速来计算 5 小时(或天)累算干燥度指数(AD5)、5 小时(或 5 天)累算湿润度指数(HD5)、最大瞬间风速超过 33 m/s 之台风的强风指数(GW7)、3 个月的干燥度指数(TMAD)以及干热风灾害指数(HI)。它们分别由公式(1)～(5)计算得出:

$$AD5_i = \sum_{j=1}^{5} MT_{i+j} / \sum_{j=1}^{5} PR_{i+j} \tag{1}$$

$$HD5_i = \sum_{j=1}^{5} PR_{i+j} / \sum_{j=1}^{5} MT_{i+j} \tag{2}$$

式中,$i=1,2,\cdots,48(60)$,且台风袭击当天凌晨 1 时(7 月 1 日)$i=1$。MT 是 1 小时(或天)的平均温度,而 PR 是 1 小时(或天)的降水量。

$$GW7_i = Guwd_i/7, \tag{3}$$

式中,i=0418 和 0613,分别发生于 2004 年和 2006 年,且 Guwd 是台风袭击期间超过 33 m/s 的最大瞬时风速。

$$TMAD_i = \sum_{j=1}^{3} MMT_{i+j} / \sum_{j=1}^{3} MPR_{i+j}, \tag{4}$$

式中,$i=1,2,\cdots,10$,在 1 月 $i=1$;MMT 是月最大温度值;MPR 是月降水量值。

$$HI = \frac{GW \times \sum_{i=1}^{5} MT_i}{30 \times \sum_{i=1}^{5} PR_i} \tag{5}$$

式中,在台风袭击日 $i=1$,MT 是日最大温度值,PR 是日降水量值,GW 是台风袭击期间的最大瞬间风速。

2.2 图像分析指数

乔灌木树种树冠垂直剖面的各种指数由图像分析法估测而来。目标树垂直剖面的图片在平地上用 CCD 数码相机(佳能 IXY6.0)拍摄,拍摄距离依树冠大小而定,以树冠刚好充满摄像屏幕为准。拍摄位置的确定依照对目标树的观察,以确保拍摄到准确的侧面像为准,然后根据下面的公式(6)计算迎风面和被风面的树冠面积比(CAP),其中迎风面和被风面的划分依主干的中央轴为参照。

$$CAP = \frac{\text{树冠迎风面像素数}}{\text{树冠被风面像素数}} \tag{6}$$

树冠绿色面积率(CGAP)即树冠绿色部分和整个树冠剖面的像素比。测定之前,依据过渡色目视将树冠绿色和非绿色部位区分开,在获取树冠绿色和非绿色部位的像素数后,用下面的公式(7)计算:

$$CGAP = \frac{100 \times \text{绿色部位像素数}}{\text{整个树冠的像素数}} \tag{7}$$

树冠迎风面和被风面之间单叶叶面积比(LAR)是同一树冠内迎风面和被风面之单叶的平均像素数之间的比值,其计算参见下面的公式(8),所用图像来自平板扫描仪(佳能 d125u2)扫描的离体叶片。叶面积由 Photoshop 软件中读取的像素来表示,在获取像素值之前首先对图像进行处理以除去叶片以外的部分。

$$LAR = \frac{\sum_{n=1}^{30} LPW_n}{\sum_{n=1}^{30} LPL_n} \tag{8}$$

式中,LPW_n是迎风面第 n 片树叶的图像像素数;LPL_n为被风面第 n 片树叶的图像像素数,$n=1,2,\cdots,30$;叶片样本是无目的击落叶片后机械选取的。

遭受灾害天气事件袭击后,偏冠变色和偏冠枯枝较常见于一些树种的树冠上[19],这种偏冠变色和非对称枯枝可由 Image-Tool 300 图像分析软件测得。活枝率(LBP)为枝条存活部位的长度与枝条总长的比值,其计算参见下面的公式(9):

$$LBP = 100 - \left(100 \times \frac{\sum_{i=1}^{n} LA_i}{\sum_{i=1}^{n} LT_i}\right) \tag{9}$$

式中,LA 是枝条活着长度,LT 是枝条总长度,n 为测定的枝条总数。

2.3 叶片水势和叶绿素(SPAD)值

叶的水分状态是在山口大学的室内自然环境下测定的。北美枫香树的叶片水势用压力室(PMS600)于 2007 年 8 月 11 日一个晴好的大风天测得,当天的 1 h 平均最大风速和最高气温分别是 7.1 m/s 和 31.7 ℃。迎风面和被风面各测定 10 叶,测定在山口大学内进行,样叶采取后立即进行测试。单叶的 SPAD 值是每叶 30 次重复的平均值,数值来自 SPAD-502 叶绿素计的无偏测定。

3. 结果与分析

3.1 灾害天气事件与北美枫香树的响应

2004～2008 年,日本山口地区的气候多变,伴随着一系列灾害天气事件的发生,如 2004 年创纪录的 10 次台风登陆日本列岛,强风伴随着少雨的台风 0613 号以及 2007 年生长季持续的干旱少雨。

据观察,在众多气象因子中,持续的干旱或少雨伴随着高温或强风常诱发树木的保护性反应。在 2004 年、2006 年和 2007 年,由于气候条件的差异,景观树表现出了不同的响应特征(见图 1)。从 2004 年、2006 年和 2007 年 10 月中旬北美枫香树的图像(见图 1b)可见,2004 年和 2006 年北美枫香树的绿色树顶与 2007 年的红或紫红的树顶形成了鲜明的对比。2006 年,北美枫香树的偏冠变色和落叶尤为显眼,这与 2006 年极高的灾害指数相对应(见图 1a)。

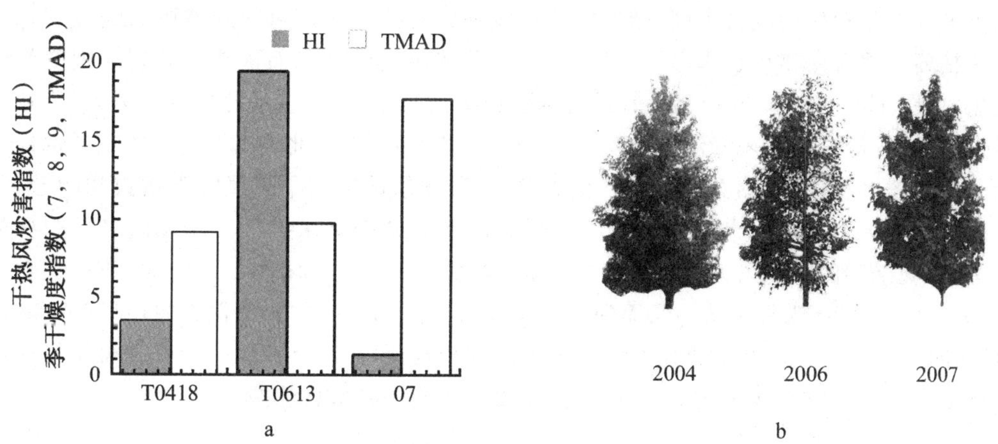

图 1 中北美枫香树在不同灾害天气事件中表现出不同的响应特征,图 a 为台风 0418 号(T0418)、台风 0613 号(T0613)以及 2007 夏季干旱(07)袭击时的干热风灾害指数(HI)和季干燥度指数(TMAD);图 b 为 2004 年、2006 年和 2007 年北美枫香树的响应特征。

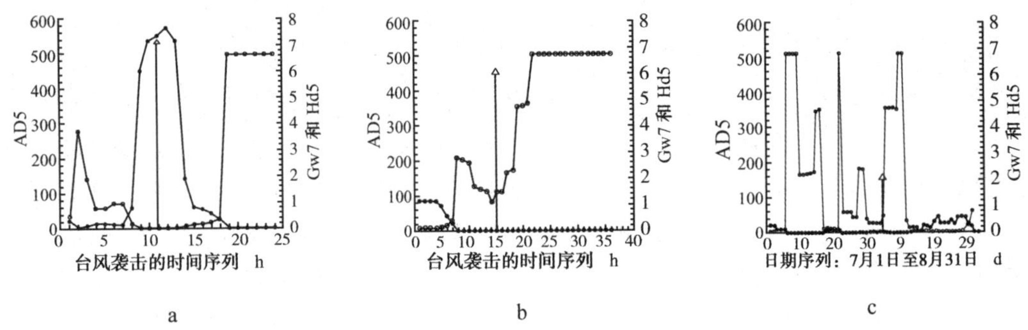

图 2 台风 0418 号、台风 0613 号和 2007 夏季干旱期间的累算干燥度指数（AD5，○-○）、累算湿润度指数（HD5，●●）和台风 0418 号、台风 0613 号和 0705 号发生期间的强风指数（GW7；标注为↑）。图 a 为台风 0418 号；图 b 为台风 0613 号；图 c 为 2007 夏季干旱和台风 0705 号事件。

2004 年，多次台风携带充足的雨水（山口市降雨量为 2224 mm）袭击了日本列岛。台风 0418 号超强的瞬间风速（山口市为 50.3 m/s）也被笼罩在了一个巨大的湿润度峰值域中（见图 2a），这明显地表现为较低的干热风灾害指数值（见图 1a），且直接导致许多景观树持续的营养生长，当年秋季北美枫香树也没有明显的树梢变红现象（见图 1b-2004）。相比之下，2006 年的前 8 个月几乎与 2004 年相似，北美枫香树持续的营养生长得益于丰沛的降水量。然而，这年 9 月中旬台风 0613 号来袭期间，无雨与强风峰值期相遇伴随着巨大的干热风灾害指数值的出现（见图 1a 和图 2b）。随后持续一个多月的干燥度峰值期无不增加了景观树的受害程度，结果使北美枫香树的树冠从迎风面到被风面偏冠变色和叶片枯萎。而且由于生长季湿润峰值突出，秋季在树冠的被风面依然未见到树梢变红的现象（见图 1b-2006）。2007 年尽管没有强台风袭击日本山口，持续的高温干旱天气，特别是在 7～9 月份众多干燥峰值的出现（见图 1a 和图 2c）诱发了北美枫香树秋季从顶部向下的非对称树叶变红（见图 1b-2007）。据观察，这种偏冠响应与干热风灾害指数相吻合（见图 1a）。与台风 0613 号相比，台风 0418 号袭击期间的强降水抵消了超强风对景观树的危害，在此期间的干热风灾害指数值也较低（见图 1a-T0418 和图 1a-T0613），所以在台风 0418 袭击山口后，未发现从迎风面到被风面偏冠变色和枯叶的北美枫香树（见图 1a 和图 1b-2004）。可以认为 2007 年夏季较高的干燥度指数值（图 1a-07）为北美枫香树树梢变红的重要诱因，而 2004 年和 2006 年的强降水抵消了夏季热波的影响。

3.2 一些景观树对台风 0613 号的偏冠响应

在 2007 年 8 月 11 日这个夏季干热多风的天气中，平均最大风速为 7.1 m/s，最高气温为 31.7 ℃。在此条件下北美枫香树的迎风面和被风面的树叶出现不同的水势（见图 3a），迎风面和被风面的水势比值（Ψ_W/Ψ_L）为 0.86。据观察，干热风袭击期间迎风面枝叶和树干遮挡效应使被风面叶片几乎处于静止状态。尽管迎风面的叶片发生萎蔫，而被风面的叶片却保持膨胀状态。两天后，随着这次大风天的结束，迎风面和被风面的水势差消失，且没有看到明显的可见症状的发生。然而，超强风伴随着少雨台风 0613 号的袭击却诱发了同一棵北美枫香树迎风面树叶的焦枯以及由此而致的偏冠变色。即使到翌年春季，树冠迎风面和被风面的树叶依然存在 SPAD 值的差异(见图 3b)。

2004～2008 年，在日本山口地区的灾害天气事件中，对景观树的伤害没有能比得上台风 0613 号的。台风 0613 号袭击之后，在许多银杏树的迎风面看到了叶焦枯现象。迎风面和被风面焦枯叶片的数量和程度的差异使得该树种表现出了明显的偏冠变色。不仅如此，从海岸到内陆也观测到了银杏树偏冠变色的区域变化(见图 3c，●●)。在台风 0613 号袭击之后，距海岸不同距离的银杏树表现出明显的枝条回枯差异(见图 3c，○○)，而枝条回枯似乎是银杏树偏冠的另一个重要原因。若依照阈值曲线趋于稳定的标准来判断，枝条回枯的银杏树主要集中在距海岸线 8 km 以内，尽管在距海岸 8 km 以外仍然可以看到一些生长在限制性立地上的、活力低下的银杏树出现抽枝死亡。图 3c 集合了树冠绿色面积率(CGAP)和活枝率(LBP)曲线，显然枝条的枯死和叶枯之间存在明显的地域差异。这意味着即使整个树冠的树叶均焦枯的银杏树翌年仍有发芽展叶的可能。

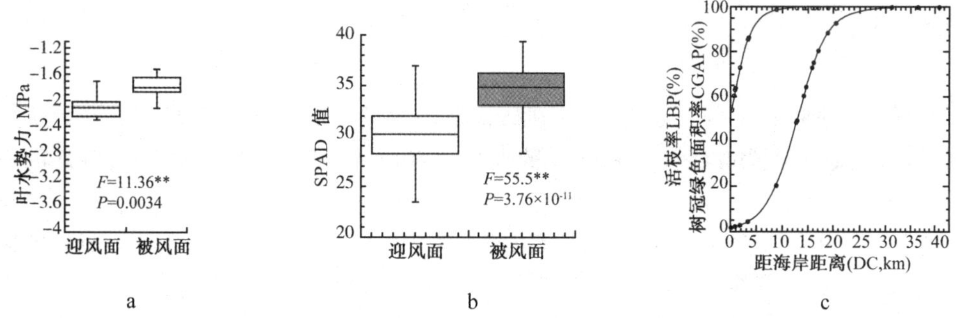

图 3 台风 0613 号等极端天气事件期间北美枫香等树种的非对称响应。图 a 为 2007 年 8 月 11 日观测到的一株北美枫香树迎风面和被风面树叶的水势差别(n=10)；图 b 为台风 0613 号袭击半年后依然观测到了该树迎风面和被风面 SPAD 值的显著差别；图 c 为台风 0613 号袭击后银杏树活枝率(LBP)与海岸距离的关系(○○)，树冠绿色面积率(CGAP)与海岸距离之间的关系(●●)。

3.3 台风 0613 号袭击后一些景观树的非对称性生长

树木生长势在很大程度上与树干、枝叶和芽的生物量累积有关，这直接反映了它们的健康状况，包括其获取水分和营养等的能力[20,21]，通常表现于不同的性状之中，如微管系统的发育程度(特别是乔灌木树种恢复期的叶面积)等。台风 0613 号袭击后变色的银杏树尽管大部分都展出了新叶，然而对其单叶叶面积比值(LAR)的观测表明，迎风面和被风面之间存在着有统计学意义的差异，甚至发生在远离海岸 10 km 以外的银杏树上(见图 4a，未修枝)。然而，这种差异在那些台风 0613 号袭击后不久进行强度修枝(仅留下一个几十年生的主干)的植株上表现并不明显(见图 4a，修枝)。北美枫香街路树在台风 0613 号袭击之前当年修枝和没有修枝(见图 4b)的植株其迎风面和被风面的树冠面积比明显地不一样。显然，由于修枝而造成的根茎比的增加为新生枝条提供了充足的水分。减少它们内部的资源竞争以至于在台风 0613 号袭击之后看不到显著的树冠非对称性。

因此，这表明相对资源的限制导致受伤严重的迎风面枝条在资源竞争中处于相对劣势，从而诱导出了它们的偏冠生长发育。

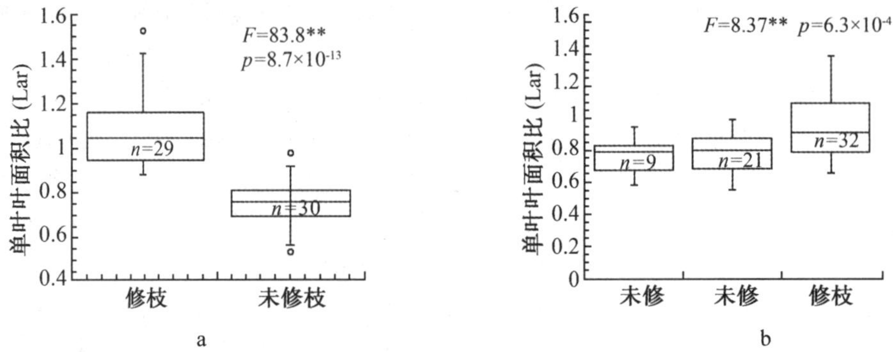

图 4　修枝与否对银杏单叶和树冠的迎风面和背风面之对称性的影响，图 a 为单叶叶面积比，图 b 为树冠面积比。

3.4　一些景观树偏冠的典型特征

与海岸边和树线上的景观树不同，在研究范围内的景观树中几乎看不到一边倒的旗形树，因为此地几乎没有盛行风和盐雾的发生。然而，在强台风等灾害天气事件的影响下，从树冠一侧到另一侧或者从顶部到基部的偏冠树不难看到，如果水平地依树干中轴和垂直地依树冠中央线为基准把树冠区分成四个象限，最典型的特征或许是树冠各象限之间的差异。由于台风的袭击和夏季热波的作用，覆盖面积最小的当属第一象限，而覆盖面积最大的则为第三象限，这当然是来自于迎风面和上部枝条的遮挡作用。图 5 所示的银杏、水杉和北美枫香树就是这种典型的景观树偏冠特征的实例。除此之外，笔者还看到一些常绿灌木树的迎风面和被风面单叶叶面积之间存在显著差异，特别是在第一和第三象限之间。在一些限制性立地条件上，甚至落叶树种上也可以看到这种不对称性。

图 5　三种落叶树偏冠的典型事例：图 a 水杉；图 b 北美枫香；图 c 银杏。

4. 讨论

灾害天气事件对树木的伤害往往通过改变树木的内在状态来实现。除了机械损伤以外，通常认为风对树木的损害在于诱导水分亏缺[6,22,23]。也有人认为盐害同样可通过诱导水分胁迫危害树木[24,25]，甚至机械摩擦也被认为是通过减少叶表皮失水的阻力形成危害的[26]。丰沛的降水不仅可以为土壤系统提供足够的补充水源，而且可以改变空气的蒸气压，减少蒸发蒸腾强度。在强台风袭击时，降水可防止植物或树木因脱水而死亡。修枝能直接降低蒸腾表面积，增加树木的根茎比，从而保持其水分的平衡。因此，修枝后的树木能承受得住强台风和夏季热波的袭击不足为奇。在许多限制性立地条件下，如石质山地、瘠薄沙地和根系生长受限地等，水分和营养的匮乏通常是树木对极端环境做出强烈反应的主要原因。增温本身也倾向于通过影响饱和水蒸气密度而增加蒸腾强度[27]，且加大水分胁迫的强度。因此，不难理解许多景观树从遭受袭击到可见的受害症状出现往往有个响应的过程[28]。

事实上，由这些因子引发的症状常具有一些共性，如叶尖焦枯、枝条回枯及器官脱落[15,17,29]等。有倾向表明，在强烈和急性的胁迫状态下，许多景观树对极端环境做出响应的顺序是从顶端到基部，如叶尖、小枝尖、树冠顶部等远离水源和营养源的部位以及低活力的部位。在极端致命的环境中，许多树种似乎都具有以牺牲局部末梢，甚至使用超敏感反应似的(HR-like)手段[16]来拯救自身的特性。在极端环境条件下，许多树木在伤及主体之前会迅速削减部分资源消耗器官或组织。

在2004年、2006～2007年，有迹象表明短期的地中海类型的高温干旱和超强台风的袭击有可能导致对景观树木的伤害，特别是那些生长在限制性立地条件下的植株，即使在山口这个降水量超过1800 mm的湿润地区也是如此。当袭击伴随着持续的干旱期时，这种伤害会更加严重。据调查，台风袭击时瞬间风速达到最大而没有降水且伴随着持续的少雨天气是诱发树木或作物焦枯的一大主要原因。

在遭受强热带风暴袭击时的自我遮挡使得树冠被风面受影响较轻。在恢复生长期的资源竞争导致树冠被风面生长更快，从而使景观树发生偏冠。但是，如果在较为适宜的环境条件下，没有后续的极端事件的影响，受害较轻之树木的迎风面能够恢复如初。据观察，景观树常接连不断地遭受灾害天气事件的袭击。经常是还没有从一次袭击之中完全恢复过来，第二次袭击便又发生了。持续的伤害导致这些景观树难以有效地防御来自生物的和非生物的侵害，从而诱使其活力下降或畸形，甚至衰老和枯死。正是这种持续地对同一植株的单向伤害导致了偏冠。

修枝，特别是全面修枝可直接增加植株的根茎比，减少器官或组织之间的资源竞争，提高景观树应对强台风的能力。事实上，在日本山口市，树木修枝业已成为当地一种流行的景观树管理方式，且造型树多种多样。这证明，修枝是一种修复偏冠和重造树形的有效方法。

主要参考文献

[1] Greulach V. A. Plant Structure and Function, The Macmillan Company[M]. New York: Collier-Macmillian Publishers, Loandon, 1973: 525-528.

[2] Kozlo M. V. Are Fast Growing Birch Leaves more Asymmetrical ? [J]. OIKOS, 2003,101: 3.

[3] Lawrance D. Some Feature of the Vegetation of the Columbia river Gorge with Special Reference to Asymmetric to Forest Trees [J]. Ecological Monographs, 1939, 9: 217-257.

[4] Noguchi Y. Deformation of Trees in Hawaii and Its Relation to Wind [J]. J. Eco. , 1979, 67: 611-618.

[5] Boyce S. G. The Salty Spray Community [J]. Eco. Monograph, 1954, 24(1): 29-67.

[6] Wardler P. Engelmann Spruce (Picea Engelmannii Engel.) at Its Upper Limits on the Front Range, Colorado [J]. Ecology, 1968, 49(3): 483-495.

[7] Van Der Valk A. G. Environmental Factors Controlling the Distribution of Forbs on Coastal Foredunes in Cape Hatteras National Seashore [J]. Canadian Journal of Botany, 1974, 52: 1057-1073.

[8] Kozlowski T. T. Water Supply and Leaf Shedding, In "Water Deficits and Plant Growth"[M](T. T. Kozlowski, 4th ed.). New York: Academic Press, 1976, 191-222.

[9] Kramer P. J. Water Relation of Plants[M]. New York: Academic Press, 1983: 187-213.

[10] Tyree M. T. , Zimmermann M. H. Xytem Structure and the Ascent of Sap, 2nd edition[M]. Berlin: Springer-Verlag, 2002: 174.

[11] Orshan G. Surface Reduction and Its Significance as a Hydroecological Dactor [J]. J. Ecology, 1954, 42: 442-444.

[12] Addicott F. T. , Lyon J. L. Physiological Ecology of Abscission, In "Shedding of Plant Parts" (T. T. Kozlowski, ed.)[M]. New York: Academic Press, 1973: 85-119.

[13] Addicott F. T. Abscission[M]. London: University of California Press, 1982: 205-207.

[14] Yapp R. H. Spiraea Ulmaria and Its Bearing on the Problem of Xeromorphy in Marsh Plants[J]. Ann. Bot. , 1912, 2: 815-870.

[15] Kozlowski T. T. Shedding of Plant Parts[M]. New York: Academic Press, 1973: 1-117.

[16] Günthardt-Goerg M. S. , Vollenweider P. Linking Stress with Macroscopic and Microscopic Leaf Response in Trees, New Diagnostic Perspectives[J]. Environ. Pollut. , 2007, 147: 467-488.

[17] Liu Y. B. , Zhang T. G. , Li X. R. , et al. Protective Mechanism of Desiccation Tolerance in *Reaumuria soongorica*: Leaf Abscission and Sucrose Accumulation in the Stem[J]. Science in China Ser C: Life Sciences, 2007, 50(1): 15-21.

[18] Millington W. F. , Chaney W. R. Shedding of Shoots and Branches, In "Shedding of Plant Parts", (T. T. Kozlowski, ed.)[M]. New York: Academic Press, 1973: 149-204.

[19] Frey B. R. , Lieffers V. J. , Hogg E. H. , et al. Predicting Landscape Patterns of Aspen Dieback: Mechanisms and Knowledge Gaps[J]. Can. J. For. Res. 2004, 34: 1379-1390.

[20] Robichaud E. , Methven I. R. Tree Vigor and Height Growth in Black Spruce[J]. Trees, 1991, 5: 158-163.

[21] Maguire D. A. , Kanaskie A. The Ratio of Live Crown Length to Sapwood Area as a Measure of Crown Sparseness[J]. Forest Science, 2002, 48(1): 93-100.

[22] Whitehead F. H. Experimental Studies of the Effect of Wind on Plant Growth and Anatomy[J]. New Phytol, 1963, 62: 80-85.

[23] Maki T. , Suzuki Y. , Kamoda F. , et al. Meteorological Disaster in Agriculture and the Countermeasure [M]. Yokendo press, 1991: 110-137.

[24] Munns R. Physiological Processes Limiting Plant Growth in Saline Soils：Some Dogmas and Hypotheses[J]. Plant，Cell and Environment. 1993，16：15-24.

[25] Pammenter N. W.，Smith V. R. The Effect of Salinity on Leaf Water Relations and Chemical Composition in the Sub-Antarctic Tussock Grass *Poa cookii* Hook F.[J]. New Phytologist，1983，94：585-594.

[26] Grace J. The Effect of Wind and a Reduced Supply of Water on the Growth and Water Relations of *Festuca arundinacea* Schreb[J]. Ann. Bot. 1982，49：217-225.

[27] Fitter A. H.，Hay R. K. M. Environmental Physiology of Plants[M]，Academic Press，2002：162-190.

[28] Rust S.，Roloff A. Acclimation of Crown Structure to Drought in Quercus Robur L.：The Abscission Zone and Its Physiological Concequnces[J]. Basic and Applied Ecology 2004，5：293-299.

[29] Treshow M. Environment and Plant Response[M]. Mcgraw-Hill Publications in the Agricultural science，1970：22-34.

（原文发表于《灾害学》2001，26(2)：5-10，30.）

第三部分　叶色变化

Persistent and advanced reddening of sweetgum leaves after major veins severing

WANG Fei

(Shandong Forestry Research Institute, 250014, Jinan, China)

1 Introduction

Morphologies of plants or trees are usually the equilibrium between genetic property and environmental effects. The catastrophically environmental extremes occasionally happen in field, even induce them into protective responses (Alexieva *et al.*, 2001; Liu *et al.*, 2007) or directly damage them (Chiba, 1994; Yamamoto *et al.*, 1996; Baig and Tranquillini, 1980). Some tree species or some individuals at special stages or status are especially sensitive to environmental extremes such as excessive irradiation. However, the excessive irradiation energy is a relative concept (Fitter and Hay, 2002). Under the condition of most environmental variables in stress status, even the scatter light may be occasionally excessive to the plants or trees. Plants or trees usually protect against the excessive photo energy acceptation through heat transformation (Donald, 2001; Fitter and Hay, 2002) and the anthocyanin accumulation in leaves to attenuate injurious irradiation energy (Chalker-Scott, 1999). Water, with larger specific heat and latent heat value of melting and evaporation (Fitter and Hay, 2002), plays an important role in energy balance of plants and trees (Chalker-Scott, 1999). The transpiration cooling effect in plant is also a procedure of excessive heat energy dissipation (Clements, 1934; Gates, 1968; Fitter and Hay, 2002). The anthocyanin content even negatively relates to leaf stomatal conductance (Farooq *et al.*, 2009), which indicates that water stress can induce the protective responses to the excessive light or heat energy.

The expanding juvenile leaves on new shoots of many tree species appear red or non-green colors, which is often called "delayed greening" (Lambers *et al.*, 1998; Numata *et al.*, 2004). The expanding or fully expanded sweetgum (*Liquidambar styraciflua L.*) leaves with the characteristic of delayed greening often appear persistent red or purplish red color at stressed part after the major veins being severed. It is named as the "persistent reddening" in the paper. The phenomenon that the chlorophyll mature leaves early become red or purplish red at stressed part after the major veins being severed is called "advanced reddening" in the study. The appearance of red top crown, persistent reddening and advanced reddening of sweetgum leaves seems an instance of transpiration cooling failure. In some extent, the elevation of anthocyanin in leaves can be considered as the result of inter-relation of many environmental factors, such as photoperiod, drought stress (Farooq *et al.*, 2009; Alexieva *et al.*, 2001; Yang *et al.*, 2005). According to the measurement of leaf stomatal conductance, thermo image temperature, water content measurement and RGB image analysis, the sweetgum leaves with major veins severed may be one of the examples of this kind of special sensitivity to the excessive irradiation energy and water stress.

2 Materials and methods

The study was carried out in Yamaguchi University, Japan. The observed sweetgum trees were several new sprouting plants from stem cutting stocks, one old stem sprouting plant and some street trees in Yamaguchi City, Japan. During the study, leaf laminas of sweetgum were usually divided into ten half-lobes in responding to their special palmate venation and in the order against hour hand. If there were seven lobes on the leaf, the smallest two lobes at leaf base were incorporated into half-lobe 1 or 10. Analysis of RGB images, thermo images and water content of leaf laminas were carried out by using these half-lobes as basic unit.

Leaf temperature was determined by using thermography (Jones, 1999; Jones *et al.*, 2002; Jones and Leinonen, 2003; Prytz *et al.*, 2003). In this study, a NEC TH7100 thermal infrared (8−14 μm) camera, with the temperature measuring range from −20 to 100 ℃ and minimum sensible temperature 0.06 ℃ centigrade, was hand-held about 50 cm above the objective leaves and then focused to clear. Thermo image temperature was measured by using the active heat method in field under direct sunshine heating from 8:00 a.m. to 10:00 a.m. Smoothly expanded leaves were selected to take the thermo images.

Leaf water content of half-lobes, which were directly cut from attached leaves and taken back to lab with plastic bags respectively, was measured by rapid weighing method in room with an electronic balance (Shimadzu Auw220, 1/10000 g) on Sep. 21st, 2009. The water contents of non-severed area (N), transitional area (T) and stressed area (S) were the average values of half-lobes in these areas.

The G/L value for half-lobes is a proportion between green and luminance value from RGB images. It was obtained with image analysis method (Wang *et al.*, 2008,

2009a, 2009b) to describe the persistent reddening and advanced reddening of leaves. Images were scanned with a scanner (Canon D125u2) or taken with a CCD camera (Canon IXY 6.0). The photo making (PM) and vein severing (VS) date were noted at figure caption, respectively.

Leaf stomata conductance from different areas on the same leaf was measured with a SC-1 leaf porometer in clear field environment from 9:00 a.m. to 11:00 a.m. on Sep. 10, 2009. During the measurement, the sensor clip was fixed on non-severed, transitional and stressed half-lobes of attached leaves and automatically measured.

3 Results and discussion

By observations, juvenile leaves from sweetgum showed varied coloration and were green or light green during leaf expansion in early April in Yamaguchi in the normal conditions. However, in growing season some sweetgum trees presented delayed greening leaves on top of the crown or twigs, especially the trees growing at constricted sites and/or during the extreme hot/dry period (Fig. 1a). As the leaves grew, the delayed greening leaves became green gradually and usually maintained two apical red leaves on shoots as reported by Hughes *et al.* (2007). The persistent reddening of expanding leaves locally appeared on leaves by carefully designed experiment in this study. Sweetgum leaves usually possess five or seven lobes on leaf lamina, which consist one main vein on each lobe (Fig. 1a). Their major-veins and sub-major-veins terminate at the end of the leaf edge respectively. There is no anastomose among major veins or sub-major veins, and only local minor veins anastomose at the joint part between two lobes (Fig. 1b). If the major vein in one lobe is severed the leaf lamina of the lobe has to obtain water only from the nearest non-severed lobes via the minor vein joints. If more than one adjacent major vein is severed, there is always a part far from these minor vein joints. Therefore, it is easy to induce the local leaf lamina far from these joints into water imbalance by major vein basally severing, for example the Ⅱ, Ⅲ and Ⅳ major veins severing (Figs. 1a, 1b), especially for the rapid expanding younger leaves. As the stress continued, persistent reddening occurred (Fig. 1b-S, 1c-S) at local area. Severed leaf could be clearly divided into non-severed area (Figs. 1b-N, 1c-N), transitional area (Fig. 1b-T, 1c-T) and stressed area (Fig. 1b-S, 1c-S). The non-severed area usually turned to green color as the leaf grew, while the stressed area maintained red color or purplish red color. The transitional area showed the intermediate coloration and more greening near the minor vein joint (Fig. 1b). The persistent reddening even persisted all of the growing season until the end of the experiment. It was observed that the major veins of Ⅱ and Ⅳ became the demarcation line between persistent reddening area and greened area (Fig. 1b, 1c) for the leaves of Ⅱ, Ⅲ and Ⅳ major veins severed. That meaned they were main barriers to interrupt water transporting into the half-lobes next to the severed veins. Therefore, they were important lines dividing the stressed area, transitional area and non-severed area. By thermograph taking, a high temperature area was observed at the farthest part from non-severed-area (Fig. 1d). From the thermo-image, the major vein barrier and the difference among non-severed area, transitional area and stressed area could also be seen. However, there was no this kind of phenomenon on normal leaves. It showed significant difference of leaf temperature before and after leaf severing.

Using the same major vein severing method, partially advanced reddening on chlorophyll mature leaves has also been induced (Fig. 2). It was clear that the advanced reddening from severed leaf (Fig. 2b) appeared significantly different color from its normal status before severing (Fig. 2a) while it was similar to that of severed leaves in persistent reddening process (Figs. 1b, 1c), which showed red color in the stressed area. The major vein barrier from vein Ⅱ and Ⅳ also appeared on the leaves Ⅱ, Ⅲ and Ⅳ major vein severed in the advanced reddening process. In the situation of major veins Ⅲ and Ⅳ severed, the major vein barriers also could be seen and showed slight leaf advanced reddening (Fig. 2c-S) at stressed area. In fact, the difference of leaf stomata conductance (Fig. 2c-L) and G/L value (Fig. 2c-G) from RGB images have been measured among the non-severed area (Fig. 2c-N), transitional area (Fig. 2c-T) and stressed area (Fig. 2c-S). The leaf similarly showed the high temperature area (Fig. 2d-S), transitional temperature area (Fig. 2d-T) and low-temperature area (Fig. 2d-N) as well as the major vein barrier from veins Ⅲ and Ⅳ 18 minutes after Ⅲ and Ⅳ major veins were severed. It was clear that the high temperature, low leaf stomata conductance (Fig. 2c-L) occurred in stressed area, which was caused by the termination of direct water supply (Fig. 2b-W).

This kind of tendency not only appeared on sunlit leaves but also occurred on shaded leaves (Fig. 2e, 2f). They also showed partially advanced reddening in stressed area on RGB image (Fig. 2e) and high temperature area in thermo-image (Fig. 2f), although the relative area was smaller than that appeared on leaves showed in Fig. 2b and it took a longer period for the symptom occurrence. However, both the persistent reddening and advanced reddening of major veins severed leaves of sweetgum indicated the characters of tree specific or age specific and so on. Fig. 2g presented an entirely reddening leaf on the main stem of a sensitive tree to abiotic or biotic stimulation, whose leaves were usually small, thin and light colored. The phenomenon of major vein barrier could still be seen on this leaf. On the other hand, a recently pruned tree with large, thick and deep green colored leaves maintained greening (Fig. 2h). By measurement, the difference of water content among the non-severed, transitional and stressed areas was very small, even no statistically meaningful difference. It matched with the phenomenon that leaves on pruned trees often turn yellow not red or purplish red in the late autumn.

Fig. 1 A persistent reddening juvenile sweetgum leaf with five major veins signed with I, Ⅱ, Ⅲ, Ⅳ and V, respectively (a, PM and VS on Jun. 1). It was severed at the base of Ⅱ, Ⅲ and Ⅳ major veins with mark "—" and divided into ten half-lobes in the order against hour hand. The same leaf (b, PM on Aug. 1) with the persistent reddening area at stressed part (S, half-lobes of 4, 5, 6 and 7), greened area at non-severed part (N, half-lobes of 1, 2, 9 and 10) and the transitional area between them (T, half-lobes of 3 and 8). Another persistent reddening stem-developed juvenile leaf (c, VS on Jun. 1 and PM on Aug. 1) and its thermograph (d, PM on Aug. 18). The persistent reddening test had been duplicated more than six leaves and these two were typical examples(See attached color Fig. 9).

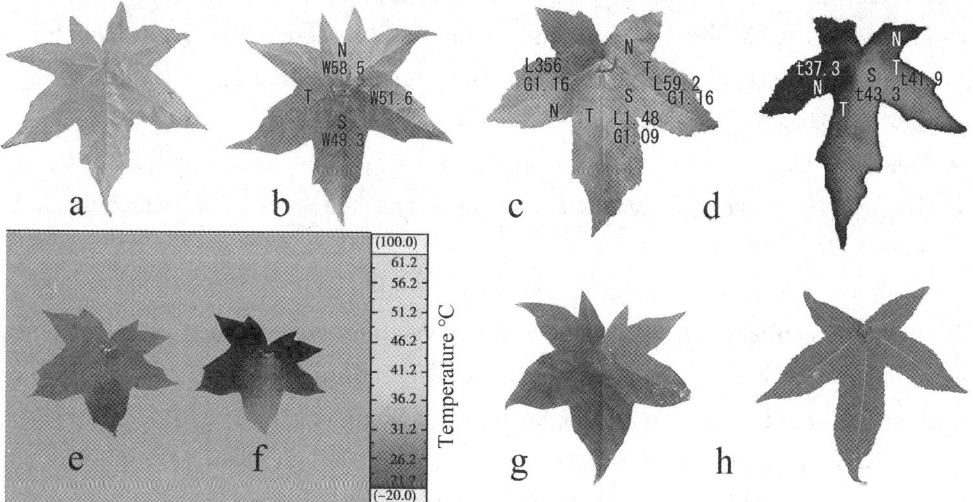

Fig. 2 A normal chlorophyll mature leaf (a, PM and VS on Aug. 9) and the stressed area advanced reddening for the same leaf after Ⅱ, Ⅲ and Ⅳ major veins were severed (b, PM on Sep. 21) with the severed positions marked with "—" and water content (W) at non-severed area (N), transitional area (T) and stressed area (S). A slightly advanced reddening leaf at the stressed area (c-S) after Ⅲ and Ⅳ major veins were severed (c, VS on Jul. 6th and PM on Sep. 6th), in which the number next to "L" is the leaf stomata conductance value and the number next to "G" is the ratio between green and luminance values from RGB image. The thermograph (d, VS on Aug. 11st and PM 18th minutes after VS) of a leaf whose Ⅲ and Ⅳ major veins were severed with different image temperatures (t). The ther-

mograph (f, PM on Aug. 18th) and the RGB image (e, VS on Jun. 1 and PM on Aug. 18th) of a shade leaf. An entire red leaf (g, VS on Aug. 9th and PM on Sep. 21th) with advanced reddening area and an entire green leaf (h, VS on Jul. 5th and PM on Sep. 4th) without advanced reddening area. Ⅱ, Ⅲ and Ⅳ major veins of leaves in e, f, g and hare h were severed. The advanced reddening test had been duplicated more than twenty times and here are some typical examples(See attached color Fig. 10).

Therefore, acute water supply reduction after major-vein was severed, especially for multi-vein-severing, caused the leaf stomata close because of the immediate reduction of leaf stomata conductance. The relative high temperature on stressed area of leaf lamina should be the direct result of transpiration cooling failure caused by stomata close and the water content reduction. The evident temperature limit from major vein indicated that it is difficult to inversely transport water into the area far from water source. This kind of barrier properly matched the limit line between persistent reddening area and greened area of severed sweetgum leaves. It indirectly suggested that there is a potential to pre-detect the water stress, persist-reddening and pre-reddening status of these sweetgum leaves by using thermography as previous researches (Chaerle and Van Der Straeten, 2000; Grant *et al.*, 2006). The persistent reddening and advanced reddening of sweetgum leaves occurred in the process of persistent water, light and heat energy imbalance after leaf severing. It is also the anthocyanin production and photoprotection mechanism in the sweetgum leaves that trigger the leaf persistent reddening and advanced reddening.

Significant contrast characteristics of climate appeared in Yamaguchi, Japan from 2006 to 2008 accompanying with some extreme weather events, such as extreme strong wind mingled with less rainfall during hit by typhoon number 13 in 2006 (T0613) and persistent high temperature and drought in 2007 (Wang *et al.*, 2009b). Prolonged vegetative growth of sweetgum trees was benefited from the abundant precipitation during the growing season in 2006 and almost showed no red top crown (Fig. 3a), although the T0613, characterized by strong wind and less rain accompanying with more than one month of no rain period, made the crown of sweetgum trees asymmetrical leaf scorching from windward to leeward. During 2007 and 2008, the persistent extreme weather of high temperature and less rainfall, especially during the growing season (Fig. 3b), induced the sweetgum trees into asymmetrical advanced discoloration from top to base of their crowns (Fig. 3a) in fall. By integration, the red top crown phenomenon was consistent with the precipitation during the growing season (Fig. 3b). It also indicated that the heavy rainfall in 2006 provided sufficient water supply to soil system, met the normal transpiration cooler requirement of trees and reduced the impact from summer heat weave. The persistent reddening and advanced reddening seemed consistent with the phenomenon of red crown top of some sweetgum trees and they may have similar mechanism.

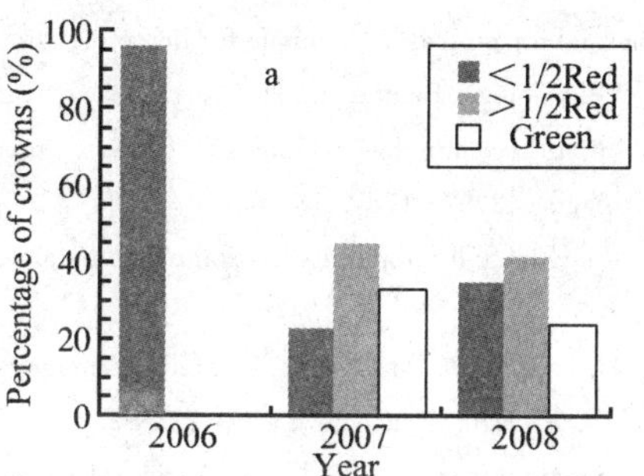

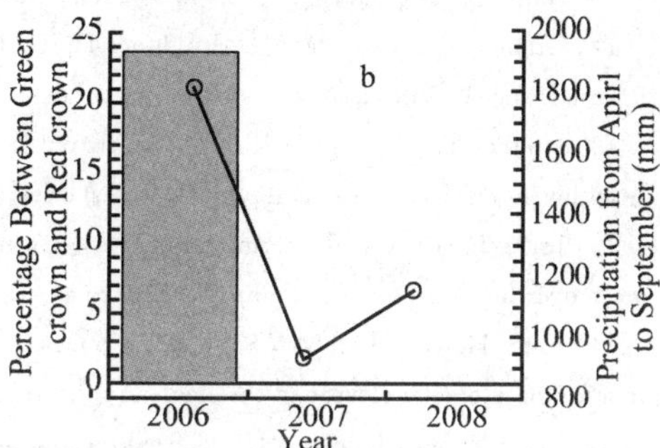

Fig. 3 The observation results of red crown top of sweetgum trees in mid-October in 2006, 2007 and 2008 (a), in which single sweet gum trees along the high way or street in Yamaguchi were visually scaled into classes of entire crown green (Green), reddening crown <1/2 (<1/2 Red) and reddening crown >1/2 (>1/2 Red). The proportion between green crowns and reddening crowns (b, histogram), and the precipitation from April to September (b, ○-○). The precipitation data was obtained from Automated Meteorological Data Acquisition System of Japan

4 Acknowledgements

The gratitude will be expressed to the Environmental Ecological Laboratory of Agricultural Faulty in Yamaguchi University, Japan for providing the experimental instruments.

References

[1] Alexieva V., Sergiev I., Mapelli S. Karanove E. The effect of drought and ultraviolet radiation on growth and stress markers in pea and wheat[J]. Plant, Cell and Environment, 2001, 24: 1337-1344.

[2] Baig M. N., Tranquillini W. The effects of wind and temperature on cuticular transpiration of Picea abies and Pinus cembra and their significance in dessication damage at the Alpine Treeline[J]. Oecologia (Berl.), 1980, 47: 252-256.

[3] Chalker-Scott L. Environmental significance of anthocyanins in plant stress responses[J]. Photochem & Photobiol, 1999, 70: 1-9.

[4] Chaerle L., Van Der Straeten D. Imaging techniques and the early detection of plant stress[J]. Trends in Plant Science, 2000, 5(11): 495-501.

[5] Chiba Y. A mechanistic analysis of devastating damage by typhoons in sugi plantations in terms of stem breaking[J]. J. Jpn. For. Soc., 1994, 76(6): 481-491.

[6] Clements H. F. Significance of transpiration[J]. Plant physiology, 1934, 9: 165-172.

[7] Fitter A. H., Hay R. K. M. Environmental physiology of plants. San Diego[M]. Tokyo: Academic Press, 2002: 57-59,131-190.

[8] Donald R. When there is too much light[J]. Plant Physiol, 2001, 125: 29-32.

[9] Farooq M., Wahid A., Basraz S. M. A. Drought stress improving water relations and gas exchange with brassinosteroids in rice under drought stress[J]. J. Agronomy & Crop Science, 2009: 368-439.

[10] Gates D. M. Transpiration and leaf temperature[J]. Ann. Rev. Plant Physiol., 1968, 19: 211-238.

[11] Grant O. M., Chaves M. M., Jones H. G. Optimizing thermal imaging as a technique for detecting stomatal closure induced by drought stress under greenhouse conditions[J]. Physiologia Plantarum, 2006, 127: 507-518.

[12] Hughes N. M., Morley C. B., Smith W. K. Coordination of anthocyanin decline and photosynthetic maturation in juvenile leaves of three deciduous tree species[J]. New Phytologist, 2007, 175: 675-685.

[13] Jones H. G. Use of thermography for quantitative studies of spatial and temporal variation of stomatal conductance over leaf surfaces[J]. Plant Cell Environ., 1999, 22: 1043-1055.

[14] Jones H. G., Stoll M., Santos T., Sousa C., Chaves M. M., Grant O. M. Use of infrared thermography for monitoring stomatal closure in the field: Application to grapevine[J]. J. Exp. Bot., 2002, 53: 2249-2260.

[15] Jones H. G., Leinonen L. Thermo imaging for the study of plants water relation[J]. J. Agric. Meteorol., 2003, 59(3): 205-217.

[16] Lambers H., Stuart C-Ⅲ F., Pons T. L. Plant Physiological Ecology[M]. New York, Berlin: Springer, 1998: 367.

[17] Liu Y. B., Zhang T. G., Li X. R., Wang G. Protective mechanism of desiccation tolerance in Reaumuria soongorica: Leaf abscission and sucrose accumulation in the stem[J]. Science in China Ser C: Life Sciences, 2007, 50(1): 15-21.

[18] Numata S., Kachi N., Okuda T., Manokaran N. Delayed greening, leaf expansion, and damage to sympatric Shorea species in a lowland rain forest[J]. J Plant Res, 2004, 117: 19-25.

[19] Prytz G., Futsaether C. M., Johnsson A. Thermography studies of the spatial and temporal variability in stomatal conductance of Avena leaves during stable and oscillatory transpiration[J]. New Phytol., 2003, 158: 249-258.

[20] Sack L., Cowan P. D., Holbrook N. M. The major veins of mesomorphic leaves revisited: Tests for conductive overload in Acer saccharum (Aceraceae) and Quercus rubra (Fagaceae)[J]. American Journal of Botany, 2003, 90(1): 32-39.

[21] Wylie R. B. Leaf structure and wound response[J]. Science, 1927, 65: 47-50.

[22] Wylie R. B. Cicatrization of foliage leaves. I. Wound responses of certain mesophytic leaves[J]. Botanical Gazette, 1930, 90: 260-278.

[23] Wang F., Yamamoto H., Ibaraki Y. Measuring leaf necrosis and chlorosis of bamboo induced by typhoon 0613 with RGB image analysis[J]. J. Forestry Research, 2008, 19(3): 225-230.

[24] Wang F., Yamamoto H., Ibaraki Y., Iwaya K., Takayama N. Evaluation Ginkgo leaf necrosis and asymmetric crown discoloration induced by Typhoon 0613 with RGB image analysis[J]. J. Agric. Meteorol., 2009a., 65(1): 27-37.

[25] Wang F., Yamamoto H., Ibaraki. Responses of some landscape trees to the drought and high temperature event during 2006 and 2007 in Yamaguchi, Japan[J]. J. Forestry Research., 2009b, 20(3): 254-260.

[26] Yamamoto H., Suzuki Y., Hayakawa S., Hirayama K. Survey on meteorological characteristics of dry summer and paddy rice damage caused by the drought in western part of Japan in 1994[J]. J. Nat. Disaster Sci., 1996, 15: 11-17. (In Japanese)

[27] Yang Y. Q., Yao Y. N., Xu G., Li C. Y. Growth and physiological responses to drought and elevated ultraviolet-B in two contrasting opulations of Hippophae rhamnoides[J]. Physiologia Plantarum, 2005, 124: 431-440.

（原文发表于 *Journal of Forestry Research*，2010，21(4)：465-468）

北美枫香主脉切断叶片的蒸腾冷却衰减和变色*

王 斐

（山东省林业科学研究院，250014，济南）

水所具有的特殊比热、溶解和蒸发潜热奠定了植物蒸腾冷却的基础，而且水分蒸腾也是植物组织最有效的散热方式之一[1]。蒸腾冷却使许多树木的温度变幅小于气温[2~6]。在阳光直射的条件下，树木通过蒸腾作用维持其温度低于受害温度值，以免遭受过量光能或热能的伤害。较低的深层土壤温度和树冠遮掩下的土壤温度本身也使得蒸腾冷却更加有效[7]。有人甚至认为，在有足够水分供应的条件下，过量的光照和高温不足以对已经适应了的植物造成伤害[8]。在水分亏缺的条件下，常观测到树木的高叶温是不足为奇的[1]。尽管叶温受很多外界条件的影响，但它常被用作反映植物水分状态的指标，因为它与叶片气孔导度[9~11]和蒸腾速率[12,13]有关。因此，与一般情况相比，水分亏缺下的高叶温被认为是蒸腾衰减的突出表现。

测定植物温度的方法多种多样，其中热红外成像法就是较直观和灵敏的方法之一。热红外成像是一种可以反映叶温[14,15]和叶片气孔导度[9~11,16,17]等异质特征的特殊工具。Prytz 等人[13]甚至用热红外成像法监测到了燕麦叶片非均质的气孔传导性和蒸腾特性。在本研究中，应用热红外成像技术检测到了主叶脉基部切断导致的北美枫香叶片局部水分亏缺和增温。结合测定叶片含水率、气孔导度等生理指标，研究了北美枫香叶基部断脉后之局部蒸腾冷却衰减和能量失衡。该部位与叶色变成红或紫红色的一致性表明，蒸腾冷却衰减诱发了断脉叶片的光保护性反应，从而使叶片达到了新的能量平衡。

1. 材料和方法

研究在日本山口大学农学部进行，研究材料来自大学校园内若干株北美枫香（*Liquidambar styraciflua L.*）树。热像温度是在 2009 年生长季节观测的，重复测定若干次，

* 基金项目：国家自然科学基金（301170671），山东省科技发展计划项目（2012GGB2201）。第一作者：王斐，日本鸟取大学农学博士，山东省林业科学研究院研究员，从事树木生态生理学、灾害气象学和光谱分析应用的研究。

每次重复 2～3 次。北美枫香叶片通常具有 5～7 个裂片，为了便于分析，在研究中一般选择具有 5 个裂片的叶片，且从右侧基部裂片开始依次编号为 1、2、3、4、5 裂片（见图 1a）。为了获得同一叶片上的水势梯度，试验设计从北美枫香叶片的上部三裂片（2、3、4 裂片）的主脉基部切断。叶脉切断的位置在图 1 中用“—”符号标记。作为对比，以同样的方式从基部切断一些山茶花（*Camellia sasanqua Thunb.*）的叶片主脉，并进行了观测研究。

红外热像用 NEC TH7100 热红外相机（波长 8～14 μm）拍摄。该相机的测温范围为 −20～100 ℃，最小感温为 0.06 ℃。该热红外成像仪对温度的感应非常灵敏，观测时应尽量避免直接接触研究对象而干扰温度测定。测定时手持相机于目标叶片上方 50 cm 处，调焦到清晰后进行拍摄。热像温度是在上午 9：00～12：00 的自然阳光主动加温过程中测得的。测定时首先选择平展的叶片顺光拍摄一张叶温较为均匀的热像，切断叶片主脉后连续拍摄热红外图像。本研究中，热像温度被定义为用 NEC TH7100 热红外相机所配备的 Thermo-workbench Ⅱ NS9200 软件读取的温度值。红外热像中一定区域的温度值是利用 Thermo-workbench Ⅱ NS9200 软件选定待测区域后直接读取的数值。然后，以 JPG 格式将红外热像存储为 RGB 图像。文中的红外热像插图使用的叶片 RGB 红外图像是用 Photoshop 软件的套索工具从热像的 RGB 图像中提取出来的。

在观测研究过程中，叶脉切断后北美枫香在切口部位会迅速溢出水分，且能嗅到挥发物质的特殊气味，3～5 分钟后切口处水分和挥发物的溢出停止且挥发殆尽，叶片有明显的阻止水分和挥发物外泄的愈合倾向。随着时间的推移，切口呈焦枯状，整个研究过程对切口未做任何处理。相比之下，切脉后山茶的切口并无明显的水分和挥发物溢出迹象，数日后切口变褐色，而且直到观测结束始终未对切口进行任何处理。

文中插图所使用的叶片 RGB 数码图像是用一台佳能 CCD 数码相机（IXY 6.0）顺光拍得的。图像的 RGB 值直接通过 Photoshop 软件读取。Green/Luminance（G/L）值由 Photoshop 得出的绿（G）和亮度（L）值计算而来[18,19]。

同一叶片不同部位的气孔导度是在室外晴天条件下用 SC-1 气孔计测得的。测定时，首先用力甩动感应头使叶室的气体与测定环境平衡，然后将其固定在半个裂片上以自动方式测定。通常情况下，每次测定应重复 10 次。

叶片裂片的相对含水率用电子天平（万分之一克岛津 Auw220 型）快速称重法测得。裂片从室外剪取之后立即放入塑料封口袋中，严密封口后待测。待所有样品采集完成后在实验室称重。在获得裂片的鲜重和干重之后，相对含水率（WC）按下面的公式（1）计算：

$$WC(\%) = \frac{F_W - D_W}{F_W} \times 100\% \tag{1}$$

式中，F_W 为鲜重，D_W 是 90 ℃时的烘干重。

叶片显微图像是用尼康 Eclipse-50i 显微照相系统拍摄的。

2. 结果与分析

2.1 北美枫香叶脉特征及其对主脉切断的反应

北美枫香具有典型的掌状叶脉，有 5～7 个裂片，每个裂片上有一条主脉(见图 1a)。其主脉和次主脉终结在叶缘，且在主脉和次主脉之间没有连接，只有裂片间结合部位的细脉之间存在连接(见图 1b)。如果某一裂片的主脉被切断，则该裂片只能通过这些细脉的连接处从未切脉的裂片获得水分。如果同时从基部切断两条以上相邻的裂片主脉，则总会有局部叶片远离这些裂片间的细脉连接处。因此，通过叶片基部叶脉切断很容易诱导北美枫香叶片的局部水分失调。尤其是那些快速伸展的幼叶(见图 2a)，因为与成熟叶相比幼叶叶片容积小，表皮蒸腾快。

相比之下，山茶叶片具有典型的羽状网脉(见图 1c)，次主脉之间在叶缘附近相互连接，而细脉在主脉和次主脉之间的间隙中相互连接。再加上叶片较厚、角质层发达以及叶表具光泽等特性，迫使山茶叶片对基部主脉切断不敏感，这主要表现在山茶 G/L 值的叶尖和叶基比值之中(见图 2c)，即从基部切断主脉的山茶浅红色幼叶与该叶片成长到绿色成熟叶时的 G/L 值之叶尖叶基比值差异不明显(见图 2c)(p=0. 11998)。这表明，基部切脉对山茶花叶片的水分输导影响不明显，水分通过次主脉或细脉连接足以满足整个叶片对水分的需求。

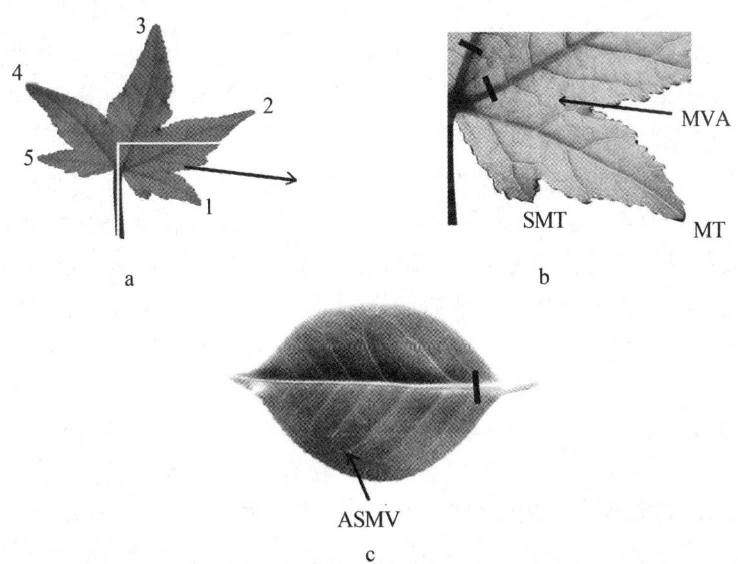

图 1　北美枫香和山茶花叶脉特征。图 a 为一片完整的五裂片北美枫香叶，每个裂片有一主脉，且从基部右侧开始标记为 1、2、3、4、5；图 b 为从北美枫香叶上分离而来的一裂片，包括其主脉终结(MT)、直达叶缘的次脉终结(SMT)以及裂片间的细脉节点(MVA)，“—”为切脉部位；图 c 为一完整肥厚肉质的、具羽状网脉的山茶花叶片，隐约可见其次主脉间(ASVV)和细脉间的连接。

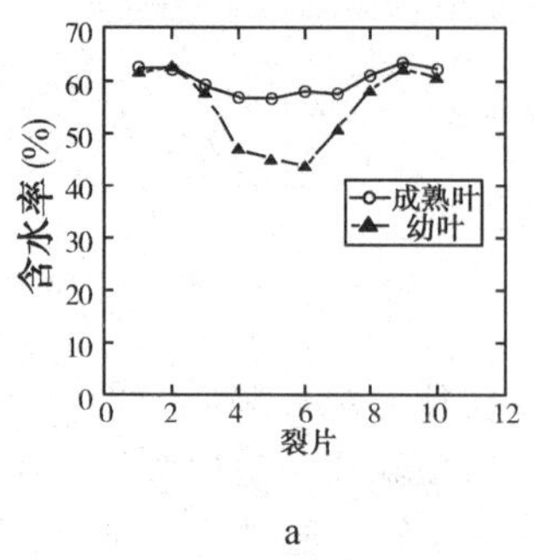

a

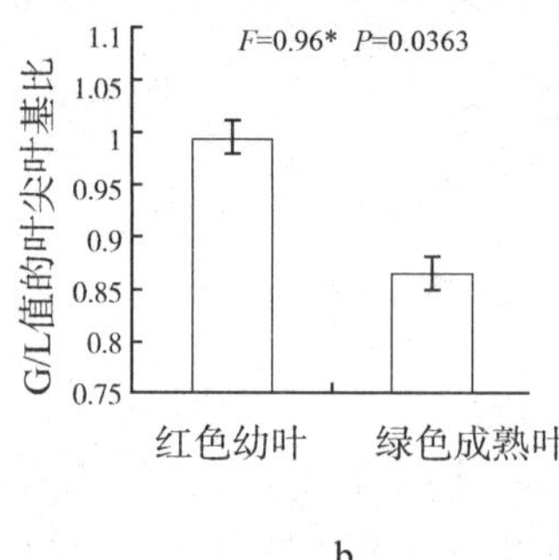

b

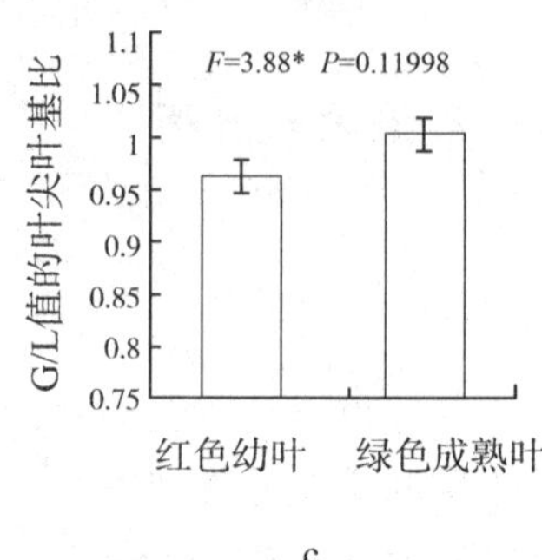

c

图2　在2、3、4裂片主脉基部切断后北美枫香成熟叶(a—○)和快速伸展的幼叶(a—▲)之半裂片含水量的对比;北美枫香叶(b)和山茶叶(c)之红色幼叶(切脉前)及其绿色成熟叶(切脉1个月后)的G/L值的叶尖叶基比。其中,山茶叶尖和叶基是从叶长轴的中间位置进行区分的;而北美枫香叶是以第2和第4主脉区分出叶尖和叶基部分的。实验重复3次。

相比之下,从基部切断2、3、4主脉的北美枫香叶片,其2和4主脉是叶片水分传输到叶片顶部的屏障。若以第2和第4主脉为分界线将叶片分成上、下两部分,则基部的G/L值明显大于顶部。其红色幼叶之G/L值的叶尖叶基比在切脉前和切脉1个月后有明显的差异(见图2b,$p=0.0363$)。这表明,具落叶属性的掌状北美枫香叶片比常绿的山茶叶片对基部主脉切断更加敏感。

2.2 北美枫香主脉切断后水分胁迫及蒸腾冷却衰减

据观测,主叶脉切断后不久北美枫香未断脉裂片(见图3a,○—○)和断脉裂片(见图3a,●●)的气孔导度均出现一个持续降低的过程。未断脉裂片和断脉裂片的气孔导度之间也存在明显的差异(见图3a),且断脉裂片的气孔导度值更小。在未断脉裂片和断脉裂片之间形成了一个水分和气孔导度的梯度。显然,较细的未断主脉不能维持整个叶片切脉前的水势和气孔导度,此时由维管系统供给的水分不足而使蒸腾衰减出现在远离水源的裂片上。断脉裂片的气孔导度可持续降低到非常低的数值,甚至难以测得。相反,未断脉裂片上的气孔导度可维持相对较高的数值(见图3a,○—○),随着时间的推移,其气孔导度可逐渐恢复到正常数值范围内,这表明较低的气孔导度来源于主脉切断而导致的水分胁迫。

从基部切断单个裂片的主脉时,断脉裂片和未断脉裂片之间的相对含水率只有很小的差异(见图3b-4)。当叶片从基部切断多个主脉时,在远离未切脉的裂片上观测到了明显的低含水率,而且水分亏缺的程度在不同的裂片上表现不一。在切断2、3、4、5裂片主脉的叶片上,裂片3和4的含水率明显低于其他裂片(1、2、5、6、7),从而处于较严重的水分胁迫状态(见图3b-2、3、4、5)。同理,切断1、2、3、4裂片主脉的叶片,裂片2和3的含水率较低(见图3b-1、2、3、4)从而表现出严重的水分失调。显然,通过不同的主脉切断组合可以获得不同水分亏缺梯度的叶片,尤其是对处于快速发育中的幼叶而言。因此,基部主脉切断可以导致北美枫香叶片局部水分胁迫和蒸腾衰减,尤其是在盛夏季节。

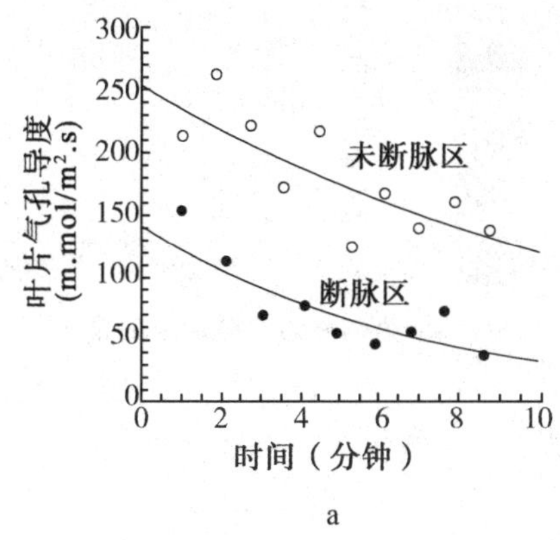

a

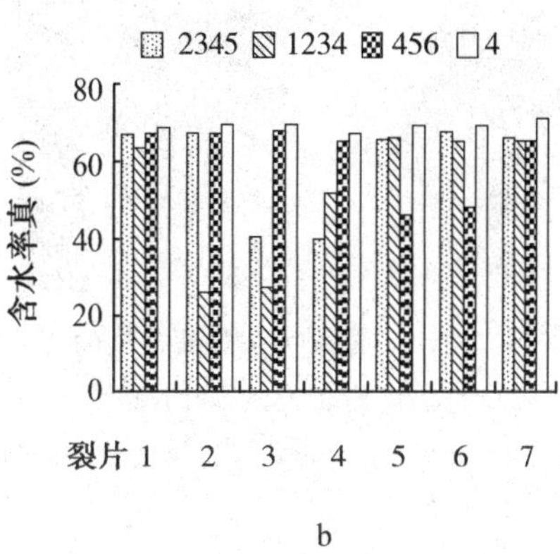

b

图3 图a为在基部主脉切断后枫香叶的断脉区域和未断脉区域的气孔导度；图b为主脉2345、1234、456以及主脉4被基部切断的枫香叶各裂片的含水率。

2.3 基部断脉的枫香绿色成熟叶局部高温、低气孔导度和变红

从基部切断2、3、4主脉的叶片上可见，断脉裂片和未断脉裂片之间的细脉连接处成为断脉裂片水分的主要来源。切脉后在远离水源的裂片上观测到了局部的高温区（见图4b），且与切脉前有显著的差别（见图4a），高温区和低温区之间的温差可达2.5 ℃左右，甚至更高。而且在2、4主脉处断脉区与未断脉区的叶温截然不同。

一些山茶叶片甚至在主脉基部切断2个月后的红外热像上仍难以观测到明显的温度差异（见图4c）。均匀的山茶叶温意味着在基部主脉切断的叶片上没有明显的水分胁迫和能量失衡的发生。与此同时，从基部切断主脉后，叶温的叶尖叶基比在山茶和北美枫香之间存在明显差异（见图4d，$p=0.02198$）。因此，大量山茶叶在主脉基部被切断半年后依然正常存活如初。

伴随着水分胁迫和蒸腾冷却衰减的持续，快速伸展期切断主脉的北美枫香幼叶之高温区常出现叶焦枯症状（见图5c）。这意味着在这一区域过量的光能和高温刺激叶片将严重受伤的部位分离出去以缩减蒸腾表面积[20]。然而，在这些特别极端的环境以外，枫香主脉基部切断1个多月甚至更长的时间内，明显地表现出叶片局部变红或变黄（见图5a），且断脉北美枫香叶的紫红色区域（见图5a）与红外热像中的高温区（见图4b）相吻合。变色部位集中在叶片横切面的上表皮和栅栏组织中，在下部海绵组织中未见红色素的存在（见图5d）。而且，这种变色与一些研究报道中涉及的光保护特征相似[20]。相比之下，山茶叶片几乎没有这种叶色的变化（见图5b）。

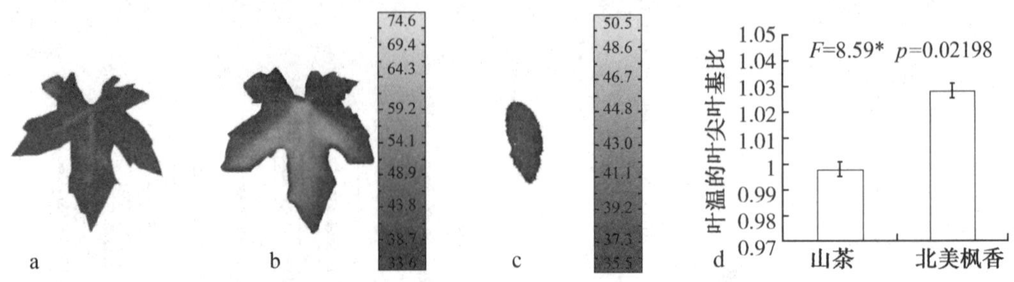

图 4. 图 a 为切脉前拍摄的北美枫香叶片红外热像；图 b 为切脉后呈现明显局部高温区的北美枫香叶片的红外热像；图 c 为叶温均匀分布的山茶切脉叶片的红外热像；图 d 为以及主脉切断的北美枫香和山茶叶片的叶尖叶基温度比。重复 4～5 次。

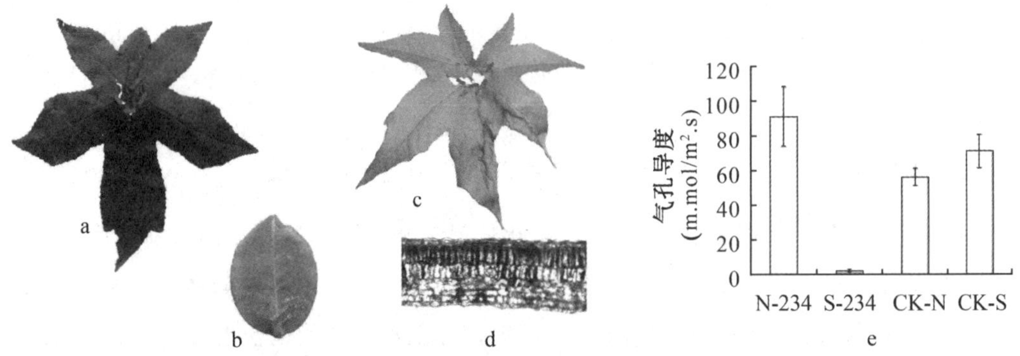

图 5. 一片 2、3、4 主脉基部切断后表现出局部变红的北美枫香绿色成熟叶(a)；一片无显著变化的山茶叶(b)；一片切断了 2、3、4、5 主脉的北美枫香嫩叶，在第 3 和第 4 裂片上出现了明显的焦枯症状(c)；北美枫香变为红或紫色之叶片的横切面显微图像(d)；一片 2、3、4 主脉基部切断的北美枫香叶绿色区域之气孔导度(e，N-234)和红色区域的气孔导度(e，S-234)，以及未切断主脉的北美枫香叶的对应部位之气孔导度(e，CK-N 和 CK-S)，气孔导度测定重复 3 次(参见彩色附图 11)。

在基部切断 2、3、4 裂片主脉之北美枫香叶片的断脉裂片上，气孔导度非常微弱且与未断脉的裂片相比存在显著差异(见图 5e，$p<0.001$)。然而，这种差异未出现在未切脉的对照叶片上，对照叶片的气孔导度甚至显示出不明显的逆趋势(见图 5e-CK)。这意味着断脉叶片上存在局部的水分胁迫和蒸腾衰减且最终诱发局部的叶色变红。

3. 讨论

不同树种具有异样的叶脉系统、表皮、气孔和叶肉组织结构，对主脉基部切断的反应同样有很大差异。落叶的北美枫香掌状叶片与常绿树种山茶花叶片的对比就非常有代表性。这也是树木对极端环境的多样适应性的一种表现[21]。充分理解和把握这些特性对于我们更好地栽培、管理和利用这些树种具有重要的意义。

作为植物或树木的一种重要组成成分，水分参与其能量代谢在很大程度上是通过蒸腾作用实现的。有许多报道发现,在水分亏缺的条件下气孔关闭且蒸腾冷却失调[10~13],这与北美枫香主叶脉切断后所测得的气孔导度结果相似。从基部切断枫香叶多条主脉后气孔导度的降低和两极分化表明,叶片对主脉切断有一个适应的过程,在水分胁迫达到一定的临界阈值后甚至会分离部分受害严重的部位而发生局部焦枯,其余部位则逐步恢复正常。

特殊的掌状叶脉和裂片间的独立性使北美枫香叶比常绿的山茶叶对基部切断主脉为更加敏感。在基部切断多个主脉时,局部高温发生在北美枫香叶片上,尤其是在阳光直射或暑热天气条件下。持续的光照加温增加了断脉区和未断脉区的叶温差异,叶温的这种镶嵌分布特征更易于用红外热像来检测。断脉叶片温度梯度的存在使得热像拍摄更容易且误差较小。同一叶片上断脉和未断脉裂片之间的可比性远远高于不同叶片热像间的绝对温度值的比较。

本研究应用红外热像测温技术测得北美枫香断脉叶片上局部的高叶温,这表明切断主脉的北美枫香叶上存在水分胁迫和蒸腾冷却的差异。北美枫香只在叶片上表皮及其相邻的栅栏组织变红且断脉后的高温区域与叶色变红的区域吻合,表明蒸腾冷却衰减诱发了其光保护过程和叶片局部的叶色变化。这意味着光保护机制易发生于持续水分胁迫或蒸腾冷却衰减的条件下。

4. 致谢

本研究是在日本山口大学农学部环境生态学研究室完成的,在此对相关的老师和同学表示感谢。

主要参考文献

[1] Fitter A. H., Hay R. K. M. Environmental Physiology of Plant[M]. San Diego, Landon: Academic Press, 2002, 231.

[2] Clements H. F. Significance of Transpiration[J]. Plant Physiology, 1934, 9: 165-172.

[3] Gates D. M. Transpiration and Leaf Temperature[J]. Annual Review of Plant Physiology, 1968, 19: 211-238.

[4] Lange O. L., Kappen L., Schulze E. D. Water and Plant Life[M]. Berlin, New York: Springer-Verlag, 1976, 143-145.

[5] Thomas M., Ranson S. L., Richardson J. A. Plant Physiology(5th edition)[M]. Landon: Longman Group Limited, 1973, 304-306.

[6] Levitt J. Responses of Plants to Environmental Stresses[M]. New York, London: Academic Press, 1972, 276-280.

[7]Rosenberg N. J. Microclimate: The Biological Environment[M]. New York, London: Jone Wiley & Sons, Inc., 1974, 68, 153.

[8]Mansfield T. A., Jones M. B. Photosynthesis: Leaf and Whole Plant Aspects, In Hall MA (Ed.)[M]. Landon, Basingstroke: Macmillan Press Ltd, 1976, 315-316.

[9]Chaerle L., Caeneghem W. V., Messens E., et al. Presymptomatic Visualization of Plant-virus Interactions

by Thermography[J]. Natural Biotechnology, 1999, 17: 813-816.

[10] Grant O. M., Chaves M. M., Jones H. G. Optimizing Thermal Imaging as a Technique for Detecting Stomatal Closure Induced by Drought Stress Under Greenhouse Conditions[J]. Physiologia Plantarum, 2006, 127: 507-518.

[11] Jones H. G. Use of Thermography for Quantitative Studies of Spatial and Temporal Variation of Stomatal Conductance over Leaf Surfaces[J]. Plant Cell Environment, 1999, 22: 1043-1055.

[12] Jones H. G., Leinonen L. Thermo Imaging for the Study of Plants Water Relation[J]. Journal Agricultural Meteorology, 2003, 59(3): 205-217.

[13]Prytz G., Futsaether C. M., Johnsson A. Thermography Studies of the Spatial and Temporal Variability in Stomatal Conductance of Avena Leaves During Stable and Oscillatory Transpiration[J]. New Phytologist, 2003, 158: 249-258.

[14] Omasa K., Takayama K. Simultaneous Measurement of Tomatal Conductance, Non-photochemical Quenching, and Photochemical Yield of Photosystem Ⅱ in Intact Leaves by Thermal and Chlorophyll Fluorescence Imaging[J]. Plant Cell Physiology, 2003, 44(12): 1290-1300.

[15] Omasa K., Tajima A., Miyasaka K. Diagnosis of Street Trees by Thermography: Zelkova Trees in Sendai City[J]. Journal Agricultural Meteorology, 1990, 45(4): 271-275.

[16] Grant O. M., Tronina L., Jones H. G., Chaves M. M. Exploring Thermal Imaging Variables for the Detection of Stress Responses in Grapevine Under Different Irrigation Regimes[J]. Journal of Experimental Botany, 2007, 58(4): 815-825.

[17]Chaerle L., Van Der Straeten D. Imaging Techniques and the Early Detection of Plant Stress[J]. Trends Plant Sci. 2000, 5(11): 495-501.

[18] Wang F., Yamamoto H., Ibaraki Y., et al. Evaluation Ginkgo Leaf Necrosis and Asymmetric Crown Discoloration Induced by Typhoon 0613 with RGB Image Analysis[J]. Journal of Agricultural Meteorology, 2009a, 65(1): 27-37.

[19] Wang F., Yamamoto H., Ibaraki Y. Transpiration Surface Reduction of Kousa Dogwood Trees During Seriously Losing Water Balance[J]. Journal of Forestry Research, 2009b, 20(4): 337-342.

[20] Hughes N. M., Morley C. B., Smith W. K. Coordination of Anthocyanin Decline and Photosynthetic Maturation in Juvenile Leaves of Three Deciduous Tree Species[J]. New Phytologist, 2007, 175: 675-685.

[21] Wang F., Yamamoto H., Ibaraki. Responses of Some Landscape Trees to the Drought and High Temperature Event During 2006 and 2007 in Yamaguchi, Japan[J]. Journal of Forestry Research, 2009c, 20(3): 254-260.

(原文发表于《北京林业大学学报》2013,35(1):72-76.)

水分失衡状态下北美枫香的叶变红

王 斐[1] 山本晴彦[2] 李小明[3] 张继权[4]

(1.山东省林业科学研究院,山东济南,250014; 2.日本山口大学农学部,日本山口县山口市,753-8515; 3.山东大学;4.东北师范大学)

1.引言

一定时期内,许多树种因其特殊的叶色、花色及其他器官的颜色而表现出特殊的观赏价值。嫩枝顶部的幼叶和落叶前的成熟叶常呈红或紫红色(Feild 等,2001)。许多树种的红色秋叶往往在表皮和栅栏组织中形成大量的花色素苷。至今人们并不清楚为什么在秋季叶片脱落之前植物会合成花色素苷。有人认为是作为预警色(Archetti,2000)与某些昆虫共进化的结果,属于生理功能的次生代谢产物(Matile,2000)或排出的毒素(Ford,1986)。叶色变红更多的被认为是环境驱动的树种适应机制,其中包括光保护(Feild 等,2001;Gould 等,1995;Hughes 等,2007)、抵御低温胁迫(Close 等,2002;Pietrini, Massacci,1998)以及用作渗透调节(Chalker-Scott,1999,2002)。尽管许多树种秋季的特殊叶色被认为是光周期(Howe 等,1995)以及营养元素代谢性转移的产物,但北美枫香的叶色变化对其生长过程中的水分失衡特别敏感。例如,在东亚季风气候的特别影响下,在日本并非所有的年份北美枫香叶都会变红(Wang 等,2009b),尤其是降水丰沛、日照减少的年份。本研究发现切脉、粗砂立地、枝叶徒长和落叶过程等诱发的持续蒸腾冷却失调和水分胁迫可以激发北美枫香的叶变红。所以,水分胁迫和蒸腾冷却失衡或许是北美枫香叶色变红的主要激发因素。

2.材料与方法

所研究的北美枫香树栽培在日本山口市一拱形街区内旧河床立地条件上,如图 1 所示,其右“拱柱”刚好位于正常的细粘土壤立地环境中,左“拱柱”位于粗沙质的低地立地环境中,而拱顶则位于二者的过渡地带。被研究的北美枫香树总计 86 株,这些街路树的 RGB 图像用佳能 CCD(IXY 6.0)数码相机于 10 月中旬在地面拍得,然后用 Photoshop 软件处理(Wang

等,2009a)。另外,用同样的方法研究了某国道两侧的北美枫香树,其中一侧是前一年刚全面修剪过枝条,另一侧未修剪。其 G(绿)和 L(亮度)值直接通过 Photoshop 软件读取,G/L 值的计算和分析参照相关文献(Wang 等,2008,2009a)进行。北美枫香树下的山茶(*Camellia sasanqua* Thunb.)灌木以同样的图像分析法进行观测。

为了研究叶片局部区域的水分胁迫,选用典型的 5 裂片北美枫香叶进行顶部 3 裂片主脉基部切断试验。切脉叶片可分为三个基本区域,即非切脉区(N)、过渡区(T)和胁迫区(S)。这些区域构成了本研究测定叶色、叶温、气孔导度和含水量等指标的基本区域。

如今,红外热像技术正越来越多地应用于植物或树木遭受胁迫的研究(Chaerle, Van Der Straeten, 2000; Grant 等, 2006; Jones, 1999; Jones 等, 2002; Jones & Leinonen, 2003)。本试验也应用了该方法。这些切脉叶片的红外热像用 NEC TH7100 热红外相机(波长 8～14 μm)拍摄,相机的测温范围为－20～100 ℃,最小感温能力为 0.06 ℃。测定时,手持摄像机于目标叶片上方 50 cm 处,调焦到清晰。红外热像的拍摄在上午 9:00～12:00的自然阳光照射加温过程中进行。选择平展的叶片先拍摄一张未切脉的对照红外热像,切断叶片主脉后连续拍摄。选择不同部位间温差最大的典型叶片进行结果分析。叶片的抽出用 Photoshop 软件的磁性套索工具完成。

同一叶片不同区域的气孔导度测定采用的是 Decagon SC-1 型气孔计,测定在晴天的野外环境中进行。测定时,探头被固定在相应的裂片上,以自动模式测定其气孔导度值,重复 10 次。以裂片为单位的相对含水量(WC)测定采用快速称重法于室内环境(53%～58%的 RH 和 20～25 ℃)下用电子天平(Shimadzu Auw 220 型,精度为万分之一克)按一定的时间间隔测得。在测定时,裂片采集后立即称重,烘干取得干重后计算含水率。北美枫香叶片的失水过程试验研究是在室内自然环境条件下进行的,试验以相等时间间隔称重完成。枝条的研究方法与此相同,而干重是在 75 ℃下烘干而得。

以钢卷尺测得的胸高直径(DBH),并按照韦林(Waring,1974)的"活力理论",参照树冠的浓密度评价其活力状态(Blanche 等,1985;McCullough, Wagner,1987)。

同一株树上的叶片按红、绿、黄和紫色分组,每组随机采样叶 20 片。叶柄的拉力应用 QIE 拉力计测定,该设备的测量精度为 0.1 kg。测定时,用左手手持叶片,右手手持拉力计并钩住叶柄基部,用力向上拉直到叶柄基部脱离或者拉断叶柄为止,此时拉力计上的应力即为叶柄的拉力。

叶片的花色素苷含量采用 1%盐酸甲醇浸提 24 小时后,用紫外可见光分光光度计在 525 nm 波长处测定光密度而得。花色素苷与叶绿素含量的比值以 525 nm 和 645 nm 波段的光密度测定值的比值(Rac 525/645)来估测。

数据的统计分析应用 Kaleidagraph 4.0 和 Excel 2003 软件来完成。

3. 结果与分析

3.1 不同立地条件和活力状态下的北美枫香叶变红

在现代城市中,用彩叶树种进行非对称的景观设计是较为常见的事情,甚至在同一街区栽培着不同的树种。图 1 就是在日本山口市的沙地梯度街区环境中北美枫香树非对称景观设计的一个实例。图中可见一拱形街道之 10 月中旬的图像，其左侧“拱柱”的树冠呈绿色，右侧“拱柱”的北美枫香树冠呈红色，而“拱顶”处的树冠呈红和绿过渡色。这种树冠叶色来自于北美枫香树上的绿色和红色叶片以及这些红绿叶片的组合。这些不同颜色树冠的 G/L 值同样也呈现相似的差异(见图 2a)，且达到了统计学上的极显著水平($p<0.01$)。右侧“拱柱”绿色树冠与其较大的 G/L 值相吻合，同样左侧“拱柱”红色的树冠也反映在较小的 G/L 值之中。树冠颜色与树冠直径的吻合意味着直径越大树叶越浓绿(见图 2b)，也达到统计学上的极显著水平($p<0.01$)。也就是说，在右侧“拱柱”上的北美枫香树冠呈绿色，胸高直径较大也更有活力。在两侧“拱柱”的北美枫香树下的山茶绿篱同样也表现出不一样的活力状态。地处左侧“拱柱”下的山茶绿篱树冠稀疏、叶小且主干低矮，明显比右侧“拱柱”下的山茶绿篱活力低下(见图 2d)。显然，在粗砂立地环境下的北美枫香树承受着较大的环境胁迫，诱发了其光保护作用的启动，从而使叶片变为红或紫红色。

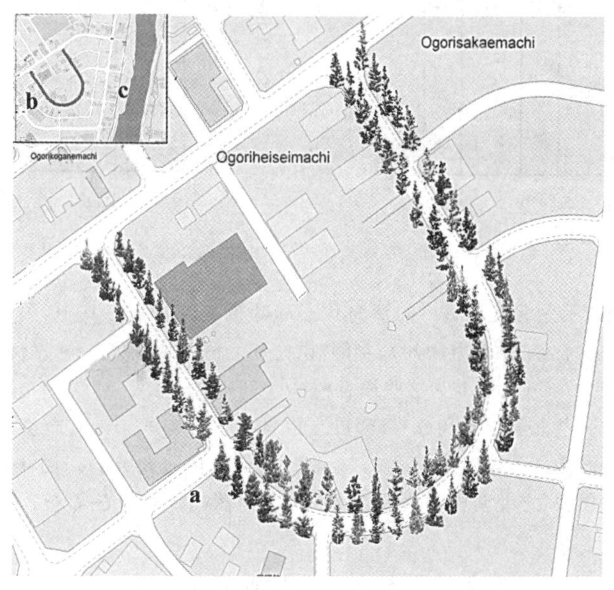

图 1　栽培在日本山口市小君之拱形街区的北美枫香树俯视图。a 为北美枫香，树冠颜色的变化反映了立地条件从粗砂到正常立地的梯度特征；左上角的小图中有红色的拱形街道 b 和靠近拱形街区之左侧“拱柱”的灰色椹野河 c(参见彩色附图 12)。

活力旺盛的北美枫香树落叶前叶色往往因叶绿素的消失而变黄色。少数植株甚至维持绿叶到翌年的春季。通常情况下，树势较弱的北美枫香叶色变红早、数量多，常呈部分树冠变红。对山口大学校园内北美枫香树之胸高直径的测定结果表明，北美枫香树的胸高直径与树冠变红的比值之间存在一种反相关。在早秋整个树冠呈绿色的北美枫香树其胸高直径也更大(见图 2c，$p<0.01$)。

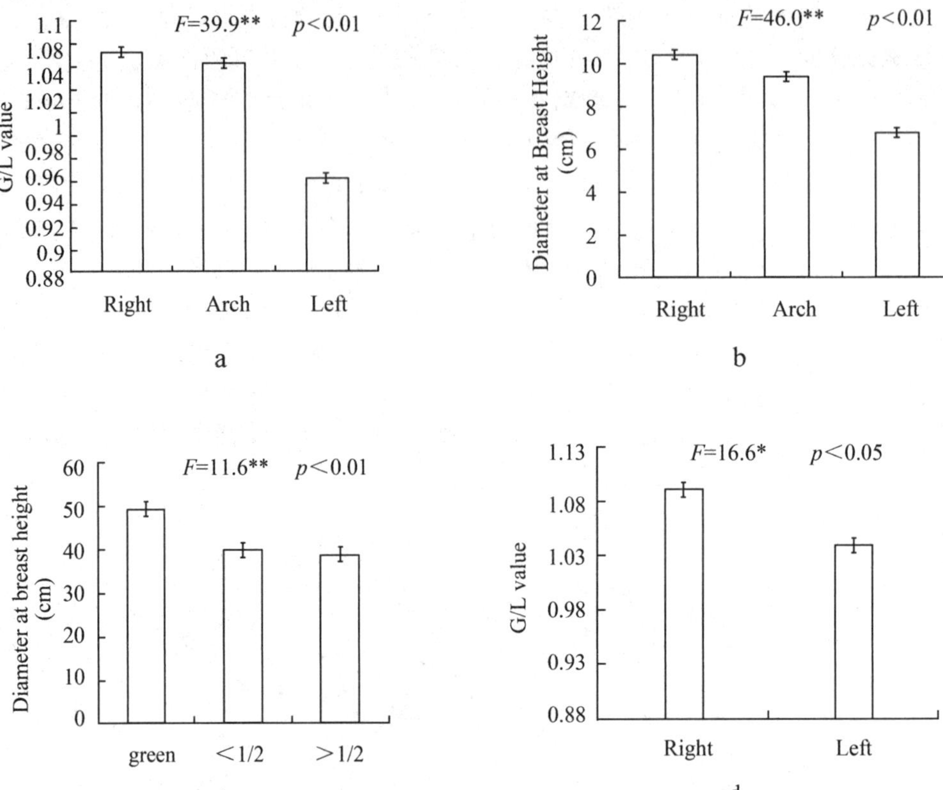

图 2 此处所示为北美枫香树的活力状态和叶色变红特征。其中，图 a 为栽培在拱形街区内的右侧“拱柱”(right)、左侧“拱柱”(left)和“拱顶”(arch)处的北美枫香树($n=86$)的 G/L 值；图 b 为这些枫香树($n=86$)的胸高直径(DBH)；图 c 为山口大学校园内一些北美枫香树($n=55$)叶变红的差异，其中“green”即整个树冠均为绿色、“<1/2”为红色树冠小于 1/2，“>1/2”为红色树冠大于 1/2；图 d 为拱形街区左侧、右侧“拱柱”之北美枫香树下栽培的山茶绿篱($n=3$)的 G/L 值。

3.2 修剪和未修剪以及主脉切断之北美枫香的叶变红

树木的地上地下生物量平衡对于其水分平衡关系至关重要。树木修剪常是调节其地上地下平衡的特殊方法，旨在维持树木较高的活力状态。在日本山口，枝条修剪在北美枫香树管理中较为常见。修剪后的植株也变红的较为少见，甚至常年维持树冠的绿色。未修剪的植株往往叶色变红早、数量多，甚至整个树冠变成红或紫色。这使得北美枫香树不仅表现出树冠变红比例的不同(见图 3a)，而且也表现出多变的树冠 G/L 值(见图 3b)。这意味着，修剪使这些枫香树能够为地上部位提供足够的水分来调节水分平衡，进而维持浓绿的树叶。

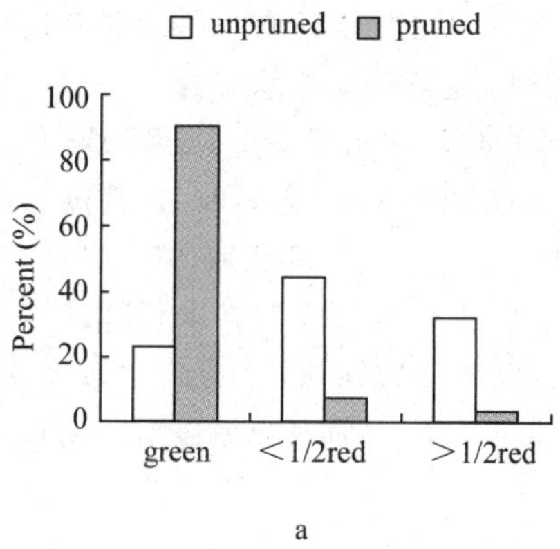

a

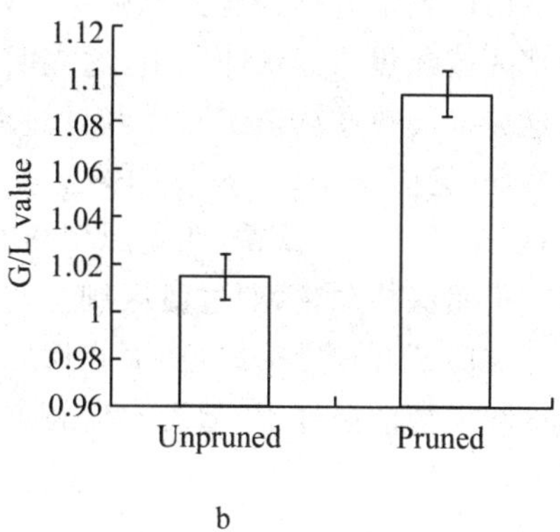

b

图 3 修剪和未修剪的北美枫香树之叶片变红的特征。图 a 为具不同树冠颜色的北美枫香树的统计结果；“Green”为整个树冠均呈绿色，“<1/2red”代表红色树冠部分小于 1/2，“>1/2red”为红色树冠部分大于 1/2。图 b 为修剪($n=19$)和未修剪($n=37$)之北美枫香树的 G/L 值。

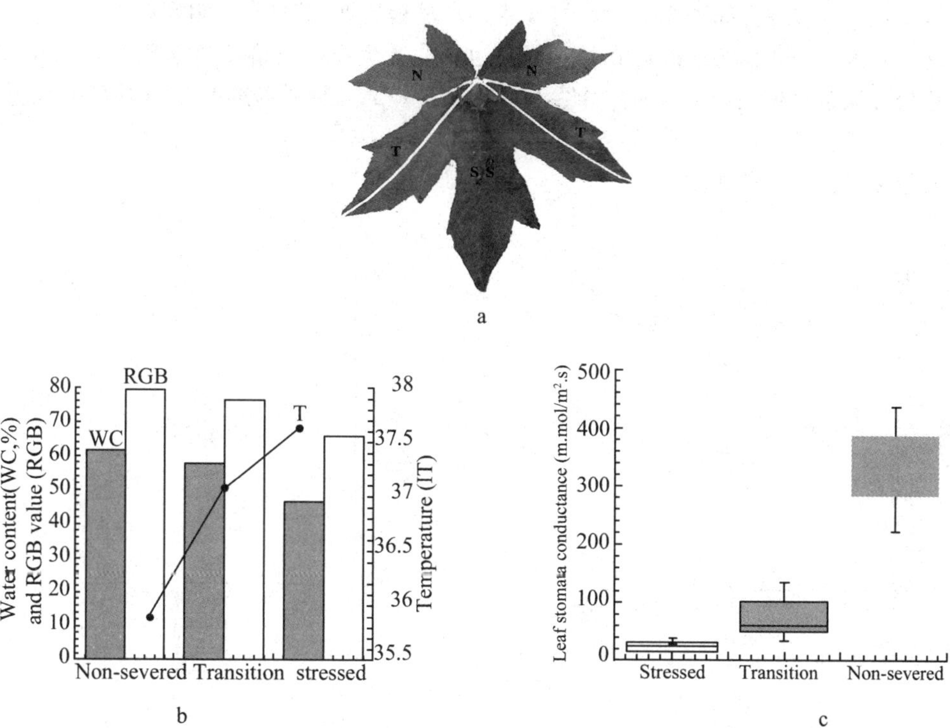

图 4 主脉切断之北美枫香叶片的水分胁迫和变红特征。图 a 为主脉基部切断的北美枫香叶的局部变红，其中包括未切脉区(N)、过渡区(T)和胁迫区(S)；叶脉切断部位由“—”表示。图 b 为未切脉区、过渡区和胁迫区的水分含量(WC)、热像温度(IT)及 G/L 值($n=3$)。图 c 为未切脉区、过渡区和胁迫区的气孔导度值($n=10$)。

在山口大学进行的北美枫香叶片切脉试验中，夏季中午时段常有一些幼叶叶片断脉后局部区域叶焦枯症状的发生，这表明基部切断主脉后北美枫香叶承受着严峻的水分胁迫。这突出反映在从未切脉区经过过渡区向切脉区的叶片含水量下降（见图 4b）和叶温升高特征（见图 4b，• — •）中，甚至表现出气孔导度的急剧降低（见图 4c）。伴随着时间的推移，尤其是到了秋季，在主脉切断的北美枫香叶片上常出现局部区域的叶变红。未切脉区域仍然保持绿色，而过渡区则呈现过渡性特征（见图 4a），这表明在未切脉区域正常的水分供应使叶片维持绿色，而在切脉的水分胁迫区叶片变成红或紫红色。也就是说，在蒸腾冷却失衡的条件下，遭受水分胁迫的区域失去能量平衡，从而使叶片变红以防接受过量的光能。

3.3 幼叶、成熟叶和不同枝条部位失水过程的差异

据观测，离体的北美枫香幼叶比成熟的功能叶失水快，尤其是在失水过程的初期（见图 5a）。这或许可以对许多树种的幼叶均表现出不同的叶色变红作出合理的解释。叶片的发育过程中，在没有防护层的特殊阶段，由于能量失衡而对水分亏缺或胁迫特别敏感。这种现象也曾发现于许多观赏树种离体叶片的自然失水过程初期，其红色幼叶的叶尖甚至比叶基失水速率要大。一些树种的幼叶在特别时期也常在叶尖产生大量花色素苷（Yapp，1912），因为水分亏缺常发生在远离水源的叶尖和叶缘部位。北美枫香枝条顶部比基部失水更快，失水率与距离枝基远近呈明显的线性正相关关系（见图 5b，$R^2=0.791$）。这与一些快速抽生的枝梢呈红或紫红色相吻合。

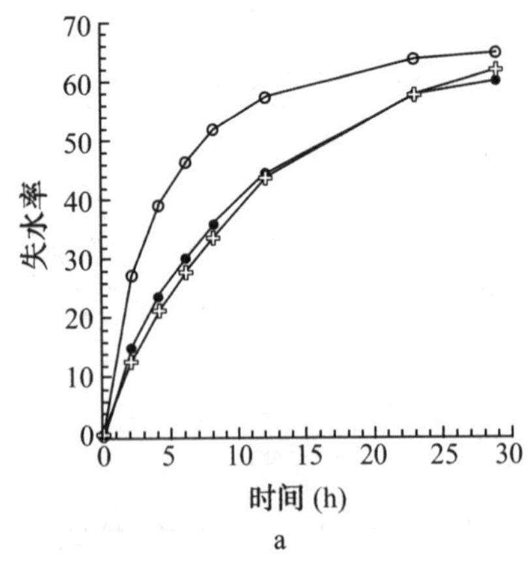

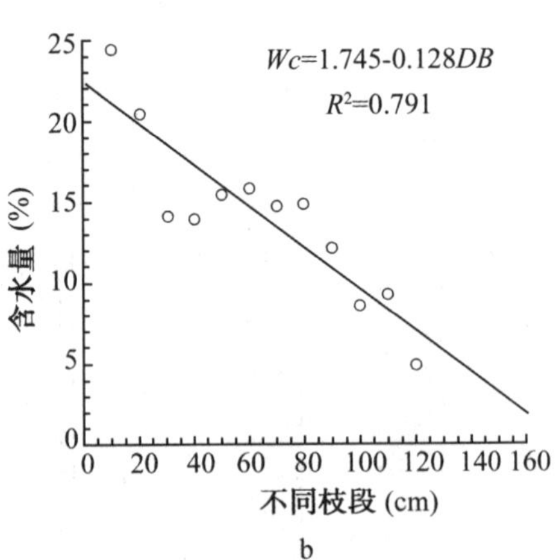

图 5　北美枫香(a)和红叶石楠(b)离体叶片失水过程。图 a 为幼叶（—○—），中等大小功能叶（—•—）和大型成熟叶（—✚—）的失水时间序列（$n=5$）；图 b 为北美枫香从基部到顶部不同枝段（长 10 cm）的失水差异，结果为这些枝段在通风干燥箱中处理 21 小时后的含水率（$n=12$）。

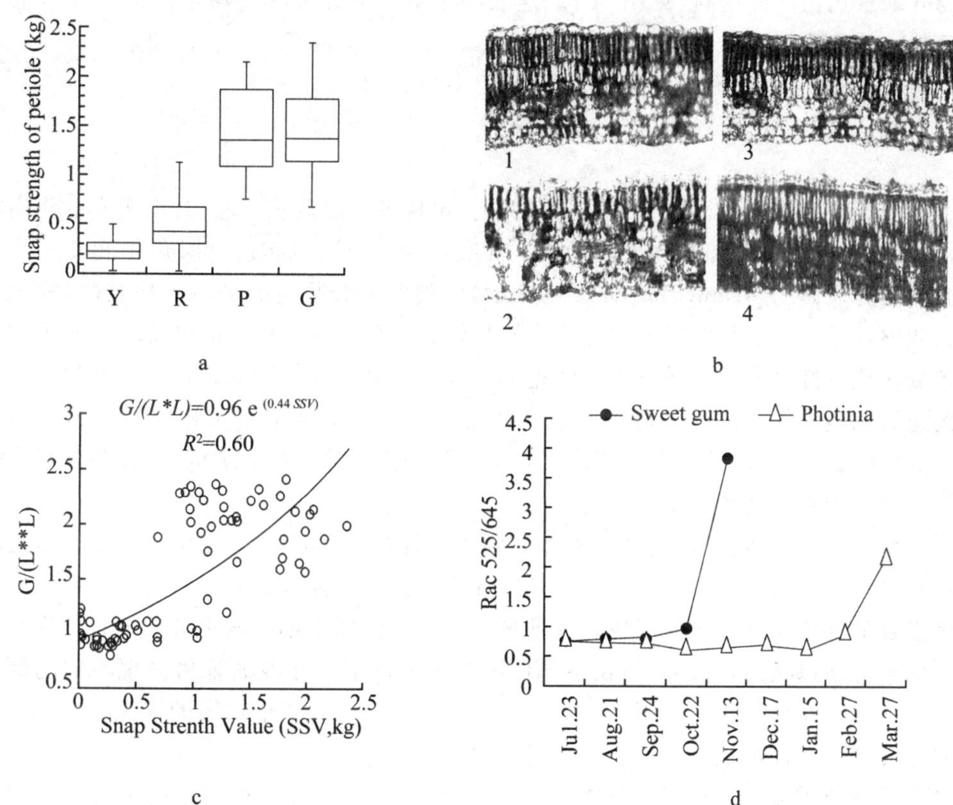

图 6 北美枫香叶色与叶柄拉力之间的相关关系以及北美枫香和红叶石楠叶片色素的季节变化。图 a 为黄(Y)、红(R)、紫(P)、绿(G)叶片的叶柄拉力值($n=75$);图 b 为红(b1)、紫(b2)、绿(b3)、黄(b4)叶片的显微横切面;图 c 为叶片 $G/L*L$ 值与叶柄拉力值之间的相关曲线($n=75$);图 d 为北美枫香和红叶石楠叶片花色素苷与叶绿素比值(由 $Rac_{525/645}$ 估测)的季节变化。

3.4 叶色变红与脱落的关系

秋季落叶前北美枫香叶片往往从上到下由紫、红、绿到黄逐渐变化。不同颜色叶片的叶柄拉力也存在显著性差异($p<0.01$)。红色和黄色叶片叶柄拉力很小或接近于零(见图 6a),以至于“一触即落”。红叶和黄叶中几乎没有叶绿素(见图 6b1 和图 6b4),只有红叶的皮下层和栅栏组织含有花色素苷(见图 6b1)。红叶和黄叶在变色后迅速脱落,这意味着叶色变化与脱落的关联。紫叶叶柄拉力与绿叶相似,在海绵组织中存在大量叶绿素,且在栅栏组织中含有大量花色素苷(见图 6b2)。秋季落叶前叶片由绿变紫的过程包括红色保护层的形成和海绵组织中叶绿素的消失。黄色叶片大多出现在树冠下部没有光保护必要的内膛之中。

叶片 $G/L*L$ 值与叶柄拉力值之间呈指数曲线关系(见图 6c,$R^2=0.60$)。在秋季落叶期间,绿色成熟功能叶与红/黄叶的叶柄拉力存在显著差异。在此脱落过程中,北美枫香叶的花色素苷与叶绿素的比值增大,而红叶石楠春季幼叶和秋季落叶前成熟叶的花色素苷与叶绿素比值同样增高(见图 6d)。这表明了落叶的北美枫香与常绿的红叶石楠之间变色的生物和生态学适应性差异。

4. 讨论

在遭受严重的水分失衡之后,不同的树种会出现各异的症状。起初,由于光合作用的低下,常表现为生长减缓和生物量的降低。其次,一些树种常通过落叶、焦枯和叶变小等来缩减其蒸腾表面积(Wang and Omasa,2012)。随着水分胁迫的进一步加剧,将发生枝条回枯或整株枯死。症状的严重程度取决于环境梯度的高低和树种的敏感性。北美枫香树叶色变化或变红是另一类适应水分胁迫的方式。主脉基部切断的裂片只好通过裂片间细脉的逆向水流从邻近的未切脉裂片获得水分。因为水分胁迫以及相应的蒸腾冷却失衡,较低的气孔导度和高叶温会迅速出现在主脉切断之裂片上的胁迫区域,尤其是在干热的夏季午间太阳直射的环境条件下。

在这些环境中,用热红外成像技术探测到了局部高温区。叶片变红区域与高温区域的高度吻合说明蒸腾冷却失衡诱发了叶片的光保护反应。并且诱发了北美枫香叶具有大量花色素苷的红色光保护层的形成。立地环境梯度导致树木不同的水分亏缺且使它们呈现不同的水分胁迫状态,所以栽培在不同环境梯度条件下的北美枫香树呈现不同的颜色是不足为奇的。在遭受干热少雨的台风 0613 号袭击后,红叶石楠树冠的迎风面出现相似的非对称叶变红或许也是这种关联性的另一事例。

作为植物或树木的一种重要组成成分,水分参与其能量代谢很大程度上是通过蒸腾过程(Fitter and Hay, 2002;Gates, 1968)实现的。经枝条修剪的个体相对较大的根系为缩减后的叶面积提供了充足的水分且保证了其正常蒸腾冷却的需要,故而能经常维持绿色叶片。一些树种的红色幼叶和枝梢与较快的失水间的一致表明幼叶和嫩枝更加需要防止过量的光照和高温胁迫。红色北美枫香树幼叶失水更快意味着叶片变红是对叶表皮变化的敏感反应。

叶片脱落涉及一个离层形成和维管束栓塞的过程(Fahn, 1990)。这些过程直接作用于叶片的水分代谢,并可诱导一些树种叶片的变红。这与许多树种叶片变红后随即脱落的现象相吻合。落叶树种秋季叶变色常被认为是光周期的日照变短、气温变低的结果。但是,许多常绿树种叶片在春末、夏初集中变红和脱落,此时的日照逐渐变长、气温逐渐升高,而脱落的叶子往往处于树冠下方接收光照较少的部位。这表明,落叶和常绿树种之间落叶和变色的方式是不同的。而常绿树种与落叶树种共同的机理似乎是成熟叶片离层的形成以及由此诱发的水分和能量失衡。水资源竞争能力低下的老龄叶片脱落以维持整株树体的水分和能量的平衡。在脱落过程中,伴随着水分输导系统的栓塞并激发光保护机制的启动。

因此,水分失衡和蒸腾冷却的失调或许是叶色变红的主要条件,这仍是需要在今后的研究中进一步证实的假说。

主要参考文献

[1]Archetti M. The Origin of Autumn Colors by Coevolution[J]. J. Theor. Biol., 2000, 205, 625-630.

[2]Blanche C. A., Hodges J. D., Nebeker T. E. A Leaf Area—Sapwood Area Ratio Developed to Rate Loblolly Pine Tree Vigor[J]. Can. J. For. Res, 1985, 15 (6): 1181-1184.

[3]Chaerle L., Van Der Straeten D. Imaging Techniques and the Early Detection of Plant Stress[J]. Trends Plant Sci, 2000, 5 (11): 495-501.

[4]Chalker-Scott L. Environmental Significance of Anthocyanins in Plant Stress Responses[J]. Photochem Photobiol, 1999, 70, 1-9.

[5]Chalker-Scott L. Do Anthocyanins Function as Osmoregulators in Leaf Tissues? [J]. Adv. Bot. Res. 37, 2002, 103-127.

[4]Close C. L., Holz G. K., Brown P. H. Effect of Shadecloth Tree Shelters on Cold-induced Photoinhibition, Foliar Anthocyanin and Growth of Eucalyptus Globulus Labill. and E. nitens (Deane and Maiden) Maiden seedlings during establishment[J]. Aust. J. Bot, 2002, 50, 15-20.

[5]Fahn A. Plant anatomy, Fourth edition[M]. Oxford: Pergamon Press, 1990, 262-263.

[6]Field T. S., Lee D. W., Holbrook N. M. Why Leaves Turn Red in Autumn: The Role of Anthocyanins in Senescing Leaves of Red-osier Dogwood[J]. Plant Physiol, 2001, 127, 566-574.

[7]Fitter A. H., Hay R. K. M. Environmental Physiology of Plants[M]. London: Academic Press, 2002, 162-190.

[8]Ford B. J. Even Plants Excrete[J]. Nature, 1986, 323, 763.

[9]Gould D. N. K., Lee D. W., Oberbauer S. F. Why Leaves are Sometimes Red[J]. Nature, 1995, 378, 241-242.

[10]Grant O. M., Chaves M. M., Jones H. G. Optimising Thermal Imaging as a Technique for Detecting Stomatal Closure Induced by Drought Stress Under Greenhouse Conditions[J]. Physiol. Plant, 2006, 127, 507-518.

[11]Howe G. T., Hackett W. P., Furnier G. R., Klevorn R. E. Photoperiodic Responses of a Northern and Southern Ecotype of Black Cottonwood[J]. Physiol. Plant. 93, 695-708.

[12]Hughes N. M., Morley C. B., Smith W. K. Coordination of Anthocyanin Decline and Photosynthetic Maturation in Juvenile Leaves of Three Deciduous Tree Species[J]. New Phytol, 1995, 175, 675-685.

[13]Jones H. G. Use of Thermography for Quantitative Studies of Spatial and Temporal Variation of Stomatal Conductance Over Leaf Surfaces[J]. Plant Cell Environ, 1999, 22, 1043-1055.

[14]Jones H. G., Leinonen L. Thermo Imaging for the Study of Plants Water Relation[J]. J. Agric. Meteorol, 2003, 59 (3): 205-217.

[15]Jones H. G., Stoll M., Santos T., et al. Use of Infrared Thermography for Monitoring Stomatal Closure in the Field: Application to Grapevine[J]. J. Exp. Bot, 2002, 53, 2249-2260.

[16]Matile P. Biochemistry of Indian Summer: Physiology of Autumn Leaf Coloration[J]. Exp. Gerontol. 35, 2000, 145-158.

[17]McCullough D. G., Wagner M. R. Evaluation of Four Techniques to Assess Vigor of Water-stressed Ponderosa Pine[J]. Can. J. For. Res, 1987, 17 (2): 138-145.

[18]Pietrini F., Massacci A. Leaf Anthocyanin Content Changes in Zea Mays L. Grown at Low Temperature: Significance for the Relationship Between Quantum Yield of PSⅡ and the Apparent Quantum Yield of CO_2 Assimilation[J]. Photosynth. Res, 1998, 58, 213-219.

[19]Prytz G., Futsaether C. M., Johnsson A. Thermography Studies of the Spatial and Temporal Variability in Stomatal Conductance of Avena Leaves During Stable and Oscillatory Transpiration[J]. New Phytol, 2003, 158, 249-

258.

[20]Wang F., Omasa K. Image Measurements of Leaf Scorches on Landscape Trees Subjected to Extreme Meteorological Event[J]. Ecol. Inf, 2012, 12, 16-22.

[21]Wang F., Yamamoto H. Detecting Leaf and Twig Temperature of Some Trees by Using Thermography Spectrosc[J]. Spectr. Anal, 2010, 30 (4): 2914-2918.

[22]Wang F., Yamamoto H., Ibaraki Y. Measuring Leaf Necrosis and Chlorosis of Bamboo Induced by Typhoon 0613 with RGB Image Analysis[J]. J. For. Res, 2008, 19 (3): 225-230.

[23]Wang F., Yamamoto H., Ibaraki Y., et al. Evaluation Ginkgo Leaf Necrosis and Asymmetric Crown Discoloration Induced by Typhoon 0613 with RGB Image Analysis[J]. J. Agric. Meteorol, 2009a, 65 (1): 27-37.

[24]Wang F., Yamamoto H., Ibaraki Y. Responses of Some Landscape Trees to the Drought and High Temperature Event During 2006 and 2007 in Yamaguchi, Japan[J]. J. For. Res, 2009b, 20 (3): 254-260.

[25]Yapp R. H. Spiraea Ulmaria L. and Its Bearing on the Problem of Xeromorphy in Marsh Plants[J]. Ann. Bot, 1912, os-26, 815-870.

（译自 *Ecological Informatics*, 2014, 19: 47-52.）

Leaf structural reddening in smoke tree and its significance*

WANG Fei, XING Shangjun, JI Yanping, YAN liping,
LIU Fangchun, DONG Yufeng, WANG Jing

(Shandong Forestry Research Academy, Jinan, China, 250014)

1 Introduction

To date, many studies have proposed mechanisms for leaf reddening that are related to photoprotection (Gould *et al.*, 1995; Feild *et al.*, 2001; Hughes *et al.*, 2007), anti-oxidation (Kytridis and Manetas, 2006), osmoregulation (Chalker-Scott, 1999, 2002), coevolution with insects (Archetti, 2000), photoperiod transition (Howe *et al.*, 1995) and low temperature induction (Close *et al.*, 2002; Pietrini and Massacci, 1998). However, because leaf reddening is a complex process that involves many factors, it has been a controversial subject for the past 100 years or more. Leaf reddening occurs in many situations, such as in juvenile leaves, old leaves, stressed leaves (Chalker-Scott, 2002) and leaves with a severed major vein (Wang, 2010). The variety of spatial and temporal conditions under which leaf reddening occurs makes it difficult to attribute the process to a single cause given that plant morphology is the result of interactions between internal metabolic processes and exogenous metamorphic actions exerted by the environment. Few studies have examined the mechanisms of heterogeneous leaf color change and even fewer have focused on the smoke tree (*Cotinus coggygria* Scop.) (Chalker-Scott, 2002) or the specific variety known as 'Royal Purple' (*Cotinus coggygria* 'Royal Purple'). In this study, we measure the water conservation ability in rapidly expanding leaves, the image temperature during leaf shedding and the structural and functional venation system of smoke trees, specifically of the 'Royal Purple'. We hypothesize that early leaf reddening in smoke trees is associated with the imbalances of both water and energy (Wang, 2013).

* Corresponding Author: WANG Fei, Email: wf-126@126.com, Tel: 0086-0531-88557773, Fax: 0086-0531-88932824. Supported by the National Natural Science Fund of China (ID Code: 31170671) and the Project of Science and Technology Development in Shandong, China (ID Number: 2012GNC11107).

2 Materials and methods

2.1 Site conditions

This study was performed between spring of 2011 and autumn of 2013 at the Yinmaquan nursery and the Yanzishan forest station in Jinan City, China, which are located at E117°03′26″ and N36°38′27″ and E117°04′57″ and N36°43′13″, respectively, as well as the site near Shandong University. Five large smoke trees of 5 and 6 meters high and approximately 20 years old, one of which was 'Royal Purple' variety and one of which was in low vigor, were used to conduct the vein severing test and snap strength assessment in 2011. Each of 15 one-year-old seedlings without transplantation (1-0) of the common smoke tree and another 10 tree/shrub species were planted at the Yinmaquan nursery and on the hill site of the Yanzishan forest station in the spring of 2012 when support was obtained from the National Natural Science Fund of China. The tree/shrub species included the Japanese black pine (*Pinus thunbergii* Parl.), Chinese arborvitea [*Platycladus orientalis* (L.) Franco], glaucous bamboo (*Phyllostachys glauca* McClure), ginkgo (*Ginkgo biloba* L.), kousa dogwood (*Cornus kousa* Buerg.), sweetgum (*Liquidambar styraciflua* L.), purple blow maple (*Acer truncatum* Bunge), Zhonghuahongye poplar (*Populus*×*euramerica* "Zhonghong"), staghorn sumac (*Rhus typhina* Nutt) and red maple (*Acer rubrum* L.). The test was designed in random block with three replicates and five plants in each block. The soil at the Yinmaquan nursery is clay loam alluvial meadow soil, while the soil at the Yanzishan site consists of rocky cinnamon soil and the soil under the street tree near Shandong University is common cinnamon soil. The soil depth at Yinmaquan nursery and near Shandong University is more than 200 cm, while the soil depth at Yanzishan is less than 20 cm. The distance from Yanzishan and Shandong University to the major meteorological station in Jinan is less than 2 km, and Yinmaquan nursery is approximately 10 km from the Yanzishan forest station. The annual mean precipitation in Jinan city is 672.7 mm, the annual mean temperature is 12.7 ℃, the mean temperature in July is 27.5 ℃ and the lowest temperature in Oct. is 0 ℃.

The number of sampled leaves and/or sampling date for the specific studies are presented in the related figure captions, e.g., the sampling number equals 11 ($Sn=11$).

2.2 Leaf vein imaging and RGB image analysis

Veins were imaged at low magnification and further examined by microphotography (Nicon Eclipse-50i). Because of the protuberant and conspicuous vein structure of smoke trees, leaf veins were photographed under direct sunlight or fluorescent lighting in spring and summer. After carefully observing the structural leaf characteristics in reddening areas, we delineated the venation system manually using the magic lasso tool (Photoshop CS2, Adobe Systems Incorporated). The RGB color image analysis of venation, leaf lamina, scorched area, stressed area and leaf base area were conducted. Unless otherwise noted, RGB images were always taken at front lighting, from the up surface of the leaves and from the sunny side of the crown.

Using the magic lasso tool in Photoshop, target leaf images were manually selected, copied and pasted onto a blank image file. Each target imaging leaf was divided into tip and base sections along the lateral vein nearest to the severed location. After selecting the tip and base sections, R (red), G (green) and L (luminance) values were recorded from the color systems. To calculate the G/L and G/R values, we followed the methods as described in Wang *et al*. (2009). We performed a contrast analysis for the RGB values between tip and base sections of sample leaves. In addition, images taken from the hill (Yanzishan) and flat sites (Yinmaquan), as well as normal and sprouting twigs, were compared using common describing statistics. The number of sampled leaves for each study is shown in the bracket of the relative figure captions.

2.3 Measurement of leaf water conservation ability and stomatal density

Leaf water conservation ability was evaluated by using the method described by Slavik (1974). Sprouting shoots were selected from seedlings (1-0) of smoke trees that had been recently planted. Three or more samples were picked from the field and then measured under natural indoor conditions: RH 50% to 70% and air temperature 25 ℃ to 35 ℃. During the transporting process from the field to the experiment site, we usually took all test materials within the same plastic bag to obtain a balance of water potential. The leaves on the sprouting shoots were classified into red juvenile or green mature leaves and then weighed with an electrical balance (Shimazhu AUY120) at specific time intervals. Water loss rate ($Wl_i\%$) was measured and calculated using Equation (1),

$$Wl_i\% = \frac{FW - Wawl_i}{FW} \times 100\% \qquad (1)$$

where FW is the fresh weight at $i=0$ and $Wawl_i$ represents the weight after persistent water loss at different time points: $i=1$, 2, 4, 6, 9, 12, 17, 24, 32, 40, 48... hours.

Current year shoots were selected to study the stomatal density according to leaf arrangement. Leaf stomatal density was directly counted from microscope views with a magnification of 200 times and five replications. This represents the number of stomata per square mm and was used to determine the developing status of leaf transpiration capacity.

2.4 Vein severing test

Vein severance was first tested on a smoke tree with deep green leaves and a tree with thin, light green leaves near Shandong University and then on all sites mentioned in Section 2. 1. More than 20 typical leaves located on the sunny side and lower part of the smoke trees were tested by cutting their major vein at the base, middle and top areas and then replicated at least three times at each site. Meanwhile, five or more replicates were performed for each vein severing pattern. Photo images were captured using a CCD camera (Canon IXY6. 0) in 2011. The images taken on Sep. 27th, Oct. 22nd, Nov. 10th, and Nov. 19th were used to analyze their coloration.

2. 5 Measurement of leaf temperature, anthocyanin and leaf stomata conductance

Leaf temperatures were measured using an infrared thermography as described by Wang and Yamamoto, 2010; Prytz *et al.*, 2003; Chaerle and Van-Der-Straeten, 2000; Jones and Leinonen, 2003; Grant *et al.*, 2007. Each 20 points were equally measured on 5 to 6 sample leaves. Leaves or leaf sections that radiated and reflected more energy than the energy they received were considered as areas with an energy imbalance. It was usually consistent with persistently abnormal high temperatures, lower stomata conductance, and leaf transpiration failure, among other factors.

Anthocyanin was extracted using methanol containing 1% hydrochloric acid over a 24-hour time period and determined by optical density (OD) readings at 525 nm with an ultraviolet and visible spectrometer (UNICO UV-2102) blanked with an extracting solution. Chlorophyll was evaluated with the same solution by the optical density readings at 663 and 645 nm. The ratio between anthocyanin and chlorophyll for 15 tree/shrub species was evaluated monthly from July 2012 to Jun 2013 by using the ratio of the OD value of the extracting solution measured at 525 nm and 645 nm. Tree/shrub species including Japanese spindle (*Euonymus japonicus* Thunb.), fortune euonymus [*Euonymus fortunei* (Turcz.) Hand.-Maz.], red leaf photina [*Photinia glabra* (Thunb.) Maxim.], glossy privet (*Ligustrum lucidum* Ait.) and the 'Royal Purple' variety of smoke tree, along with the other 10 tree/shrub species mentioned in Section 2. 1 were studied. Leaf samples from 8 deciduous tree/shrub species and 3 evergreen tree species were captured from the Yinmaquan nursery. The other 4 evergreen tree/shrub species were selected from the Shandong Forestry Research Academy, which was near the Yanzishan forest station. The relationship between water loss percent from isolated leaves in a specific time interval and leaf anthocyanin content was then regressed.

Leaf stomata conductance was measured with a portable photosynthetic system (LCi, U. K.) that had a chamber area of 6. 25 cm^2 at the time points between 10:00 a. m. to 12:00 a. m. on clear days beginning in August and then compared with the measurement of other tree species such as sweetgum (Wang, 2013; Wang *et al.*, 2014). Leaves on the sunny side of the newly planted seedlings were divided into tip area and base area, as mentioned in Section 2. 2. Measurements for each area were then recorded in triplicate by placing the chamber on the tip and base areas.

2.6 Leaf snap strength

Before leaves fell in autumn, we classified the color of leaves on each smoke tree as red, yellow, green and light green. Snap strength was measured with a QIE tension meter on a random subset of 20 samples of each color category on the trees near Shandong University. To test snap strength, the tension meter was hooked to the base of the petiole with one hand. With the other hand, the petiole was then pulled until it either detached from the leaf base or snapped into two sections. The pulling force for each leaf was represented by the instant force recorded by the strength meter.

2.7 Leaf reddening comparison in different growing status

To study leaf reddening from the samples with different growing status, several comparisons were conducted. Leaf images from newly planted seedlings at both the dry hill site and the moister flat site and from current year sprouting shoots of root stocks and normal branches of mature trees were analyzed using the RGB analysis method. These images were taken at the same period and at similar parts of the crown. In autumn, leaf samples from tip and base of the same branch on newly planted seedlings were used to evaluate the rate between anthocyanin and chlorophyll by $OD_{525/645}$.

2.8 Data statistics and graphs

We analyzed the data in Excel 2003 with the method of common descriptive statistics and used the Kalaidagraph 4.0 Software for the regression analysis and graph drawing. In the graphs, the symbol " * * " represents a significant difference among samples with a probability of more than 99%, and the symbol " * " indicates a significant difference among samples with a probability of more than 95% after the Fisher's exact test.

3 Results and analysis

3.1 Leaf reddening associating with venation

The 1° and 2° veins (Figs. 1a-1, 1a-2) within the pitch pinnate venation system of smoke trees normally ended and bifurcated at the leaf edge without network connection. The connection between 3° veins (Figs. 1a-3, 1b-3 and 1d-3) occurred only at the thin end and visible lacuna was present (Fig. 1d) in the leaf lamina. As in most other tree species, the veins of smoke tree leaves thinned gradually as they progressed from base to tip. In particular, there was almost no sign of 3° veins at the leaf edge. Leaf tips and edges often showed delayed greening (Fig. 1b) and areas with thin veins as well as interstitial areas maintained a red color longer (Fig. 1c). This characteristic can be better appreciated under an optical microscope (Fig. 1d), and accordingly, we found significant differences in the coloration patterns within single juvenile leaves as well as between light green venation and red leaf lamina (Figs. 1b, 1c).

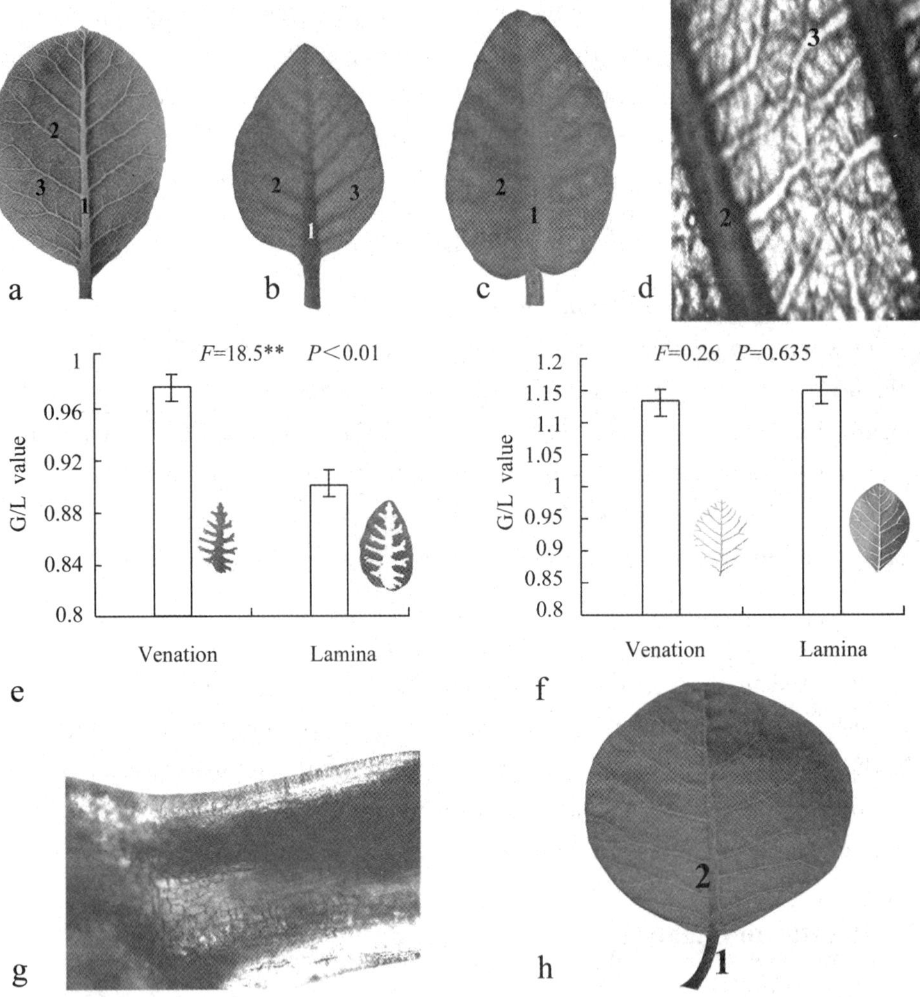

Fig. 1 Leaf venation and leaf reddening characteristics of smoke tree: (a) back surface of a normal green leaf and its pinnate venation system where 1, 2 and 3 mark the 1°, 2° and 3° veins, respectively; (b) back surface of a juvenile leaf with red tip and margins; (c) surface of typical juvenile leaf with clear red interstitial areas; (d) micrograph of juvenile leaf at early stage of reddening and with vein lacuna; (e) RGB analyzing results between venation and lamina of reddening juvenile leaves ($Sn=9$); (f) RGB analyzing results between venation and lamina of green juvenile leaves ($Sn=9$); (g) micrograph of a smoke tree leaf cross section with typical red epidermis; (h) mature leaf sample with green lamina and red petiole.

There were significantly different G/L values between lamina and venation among red juvenile leaves (Fig. 1e, $p<0.01$), which was usually found on sprouting shoots, while there were no significant differences between leaf lamina and venation system among green juvenile leaves (Fig. 1f, $p=0.635$), which often appeared on the last year twigs. We found that both juvenile and mature leaves showed early reddening at the tips, edges and interstitial veins with green color surrounding only the vascular system. These results indicate that the green and red color rations were related to water transport in the smoke tree. In other words, the local variation in leaf color was the result of heterogeneity in the vascular system. Leaf areas that were far from the venation system showed a red layer early on. We found that as the leaves grew and the lamina turned green, the major vein was further differentiated at its base and a red layer formed at the top of its epidermis (Fig. 1g, 1h-2) because there was no proper stomatal transpiration development. This process caused the red petiole to appear to extend into the leaf lamina (Fig. 1h-1).

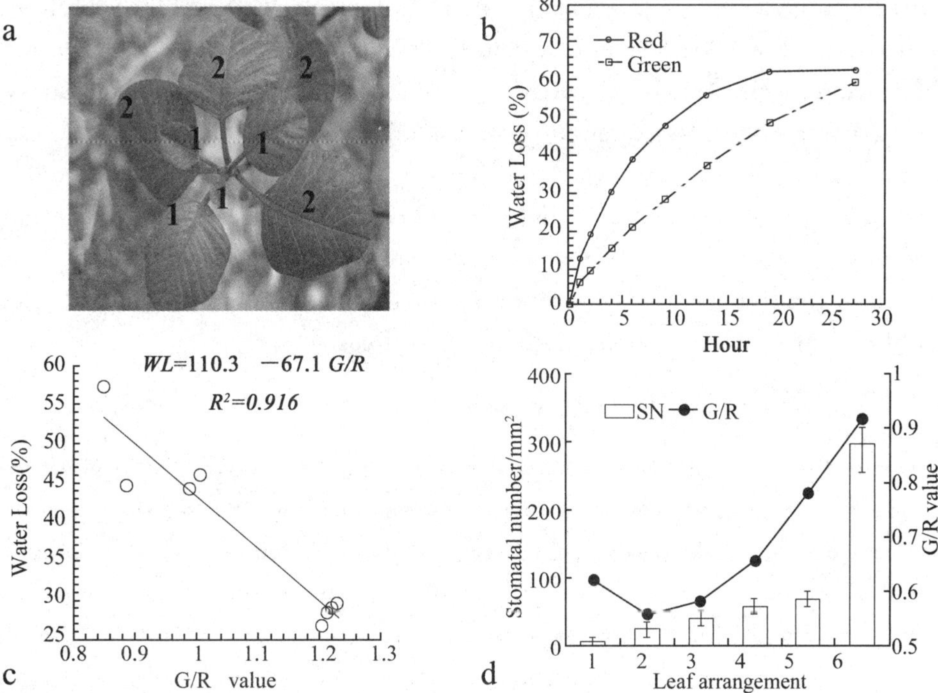

Fig. 2 Relation between water loss and leaf reddening of smoke tree. (a) leaf gradient characteristics of gradual greening from top to bottom of new sprouting shoots with leaves classified into juvenile (1) and mature (2); (b) comparison of water loss in juvenile (red) and mature (green) leaves over time; (c) relationship between water loss speed and G/R values; (d) similar tendency of leaf stomata number per mm^2 (SN, $Sn=5$) and G/R values ($Sn=5$) according to leaf arrangement.

3.2 Reddening of juvenile leaves

Juvenile leaves on sprouting shoots of many tropical and some temperate trees are often red in color. Smoke tree leaves present similar characteristics, especially in newly planted stocks. Fig. 2a illustrates similar light green venation systems and the changes in leaf color from red to green that are related to leaf development. As leaves grow, they gradually turn green from bottom to top in leaf arrangement. In this study, leaves on sprouting shoots were classified into two groups, red juvenile (Fig. 2a-1) and green mature (Fig. 2a-2). The rate of water loss in isolated red juvenile leaves (Fig. 2b-Red) was faster than that of the water loss in green mature leaves (Fig. 2b-Green), especially during the first period of tests in the spring. The rate of water loss from red juvenile leaves was exponential in shape and had a nearly linear shape in green mature leaves. Over time, the difference between water loss rates in a specific time interval became smaller until the leaves were completely dry. As a result, G/R values were negatively correlated with the water loss rate of detached leaves after nine hours of exposure with the correlation coefficients of R^2 being 0.916 (Fig. 2c). This correlation indicates, albeit indirectly, that there was a causal relationship between red color and faster transpiration from the immature cuticle layer in juvenile leaves. Furthermore, juvenile leaves showed low stomatal density until they turned green. A similar increasing tendency was found for both stomatal density and the G/R value of the leaves (Fig. 2d). This suggests that the reddening of juvenile leaves appears to be related to the imperfect development of leaf structure and cuticle layer of smoke trees.

3.3 Leaf reddening after vein severing

Due to their poor ability to conserve water, leaves of smoke tree lose water easily and quickly. The lack of network structure in leaf venation increases their sensitivity to the effects of major vein severance. Water imbalance within leaves often occurs after major vein severance, while different water gradients are induced depending on the type of vein severing pattern. When major veins were severed at the "m" and "n" points (Fig. 3a) of a leaf, the minor vein connection areas among the 2° and 3° veins near the severed location experienced a severe water stress that in turn led to water and energy imbalances near the top of the leaf. On Sep. 24th, 2011, the major or lateral veins of leaves that directly faced sunlight were severed at their base, middle and top regions. These leaves belonged to a single smoke tree with deep-green lamina. As a result, leaf scorch appeared on many leaves, especially on the second day after severing on those that were severed at the base, while some leaves with deep green and thick lamina showed no significant abnormalities (Fig. 3a). However, one month later, local scorch, discoloration and yellow spots occurred at the tops of the leaves and in the areas above the severed locations (Fig. 3b). Twenty days later, these same areas acquired orange-yellow or orange-red colors (Fig. 3c). By Nov. 19th, 2011, entire leaves had turned yellow, and some leaves that were severed at their bases showed red color on their leaf tops (Fig. 3d). From first severance to the occurrence of partial leaf reddening on normal smoke trees, it took at least two months or longer. The discolored area was only located in a fan-like area above the severed location. The effects of leaf vein severance on leaf reddening were not immediate for smoke trees that were actively growing. Instead, the severing effects on reddening became evident several weeks to two months after severance, depending on the severity of the environment. In the interaction between leaf metabolism and stress environment, advanced leaf reddening at the severed area did not occur until early autumn (Fig. 3). Therefore, smoke trees belong to the group of tree species that are known to change leaf colors during the leaf shedding process in autumn.

As previously mentioned, leaf laminas with deep-green coloration in actively growing smoke trees often scorched at the top within the first two days after severance, especially under hot, dry conditions. Under such conditions, a severe water stress and energy imbalance appeared in a fan-shaped area of the leaf to reduce the transpiration surface area and maintain water balance. As a consequence, leaf lamina were partially scorched (Fig. 3e). Once scorching occurred, the persistent imbalance near the edge of the scorched area resulted in a color change from normal yellow to red tones, such as orange-red and bright-red (Fig. 3e). In areas that were below the severed locations, leaf colors were usually normal (Fig. 3e). Leaves varied significantly in color in these areas when the tops were severed and even scorched in the patterns similar to those severed at lower parts (Fig. 3e). However, most leaves showed no scorching and only partial areas presented an orange-red color before abscission (Figs. 3f and 3g). This type of leaf color variation also occurred on leaf laminas when their minor veins had been severed at the middle and basal parts (Fig. 3g). Meanwhile, vigorous growing leaves usually turned yellow in the process of leaf abscission because they were not subjected to any real water stress or high temperatures.

Unlike leaves that only turned yellow prior to shedding, leaves on smoke trees with lower vigor, such as old or biotically stressed trees, were more sensitive to site conditions and reddened early (Fig. 3h). If leaf veins were severed during the period of color change, with most leaves being green and only a few turning to red at the twig base (Oct. 8th, 2011), leaf red color intensity was usually increased at the leaf tip. Two weeks later (Oct. 22th, 2011), referring to the severed location, leaf reddening appeared at the top area on the severed leaves, while a normal green color was maintained at the basal area (Fig. 3h).

To avoid inaccuracies in naked-eye assessments, RGB image analyses were conducted on normal leaves and color changing leaves that were severed at the base of their major vein. G/L values were the highest in the base area of major vein severed leaves (Fig. 3i-Base), intermediate in scorched areas (Fig. 3i-Scorch) and the lowest in stressed areas (Fig. 3i-Stress), with the leaves showing a red color in the stressed area (Fig. 3e). The tip-to-base ratios of G/R values were significantly less than 1.0 (Fig. 3j) ($p<0.01$).

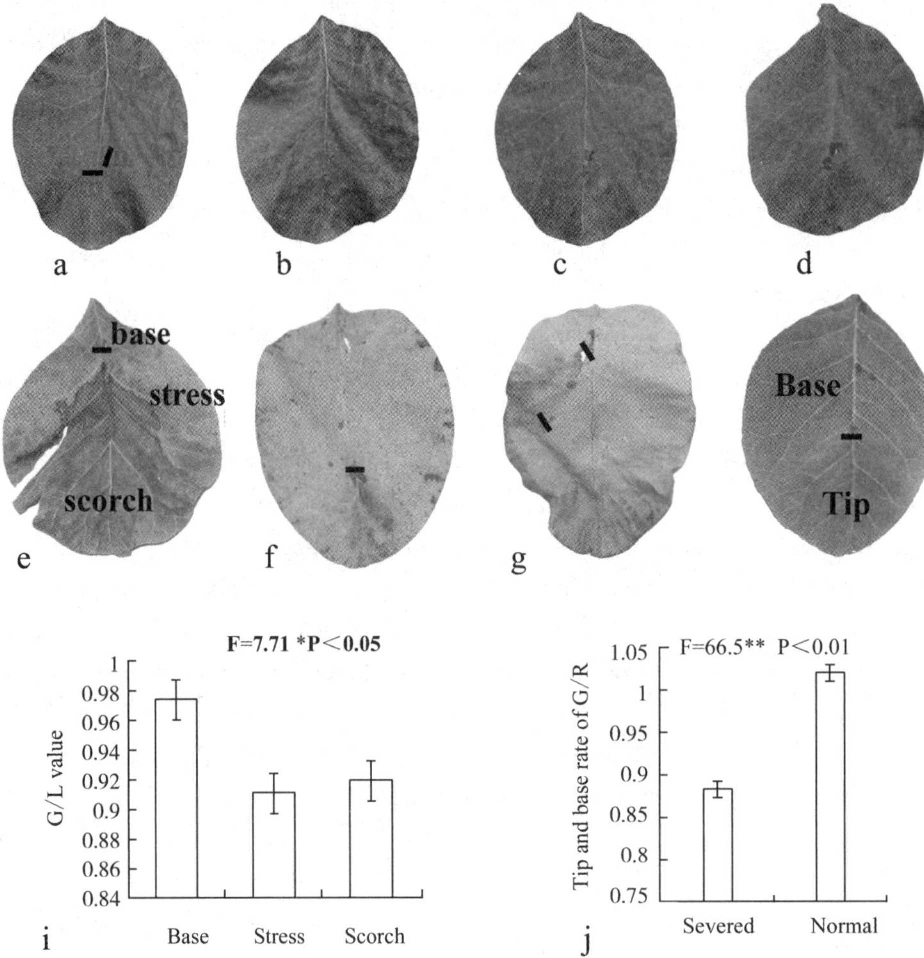

Fig. 3 Results of leaf vein severing marked with "—" and leaf RGB image analyses of smoke tree: (a) deep green leaf with severed top-medial vein m and adjacent lateral vein n on Sep. 24th, 2011 and its first image taken on Sep. 27th; (b) image taken of leaf (a) on Oct. 22nd; (c) image taken of leaf (a) on Nov. 10th; (d) image taken of leaf (a) on Nov. 19th; (e) leaf with severed major vein with a scorch, stress and base area; (f) leaf image with a severed top-medial vein; (g) leaf image with severed base and lateral veins; (h) leaf image with color difference between tip and base areas two weeks after being severed at the medial part of the major vein; (i) and (j) RGB analysis results for leaves severed at their major veins ($Sn=11$).

3.4 Persistent and advanced reddening of severed 'Royal Purple' smoke tree leaves

As previously mentioned, even the juvenile leaves of normal smoke trees that grow under favorable conditions usually show green colors in spring and turn yellow in autumn. However, the 'Royal Purple' variety usually has red or purple-red juvenile leaves even in rainy summer conditions. The 'Royal Purple' is especially sensitive to severing at the major leaf vein. Therefore, it is easy to study the color change after severing the red juvenile leaves of the 'Royal Purple'. We found significant differences between the tip and base of 'Royal Purple' juvenile leaves two weeks after severance (Fig. 4a) when severance was performed at locations between the base and middle vein. These leaves showed a persistent red leaf tip and green leaf base relative to the severing location. Normal leaves turn green in their entirety to preserve a tip-to-base ratio of G/R values near 1.0 (Fig. 4a). By comparison, both the front and back leaf surfaces of severed leaves showed persistent reddening or delayed greening at their tips because the back surface of the leaves was often turned up by wind. The tip-to-base ratios of G/R values for both front and back surfaces were less than 1.0 (Fig. 4a) and were significantly different comparing to normal leaves ($p<0.01$). Similarly, mature greening 'Royal Purple' leaves showed advanced reddening in summer (July 16th), which corresponded to approximately two months post-severance, and the difference in G/R values between tip and base areas was significant ($p<0.01$) (Fig. 4b and 4e). In addition, we measured a difference of 2 ℃ or more in temperatures between the tip and base areas of severed leaves in August (Fig. 4d and 4f). This also corresponded with an increase in leaf stomata conductance (Fig. 4c) ($p<0.01$). Temperature differences between tip and base areas with respect to stomata conductance initiated within one hour from vein severance and persisted until leaf reddening (Fig. 4d). This result suggests that in the 'Royal Purple' variety, the leaves are the most sensitive to vein severance and that persistent water and energy imbalances induce leaf reddening and partial scorching at leaf tips after major vein severance.

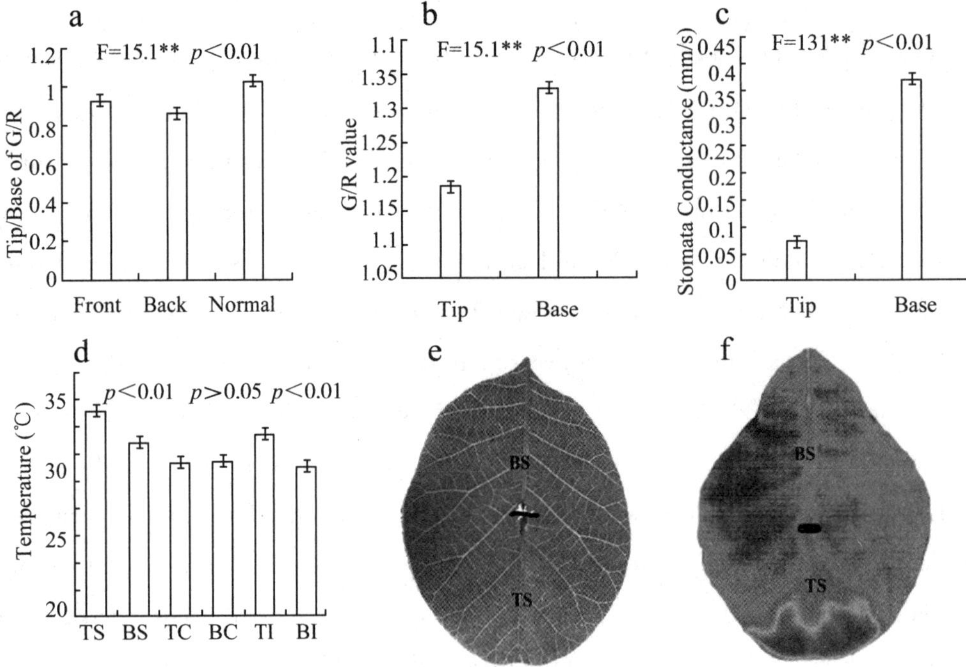

Fig. 4 Measurement results of image temperatures, stomata conductance and RGB analysis of several severed smoke tree leaves. (a) tip-to-base ratio of G/R values from both front and back surfaces of severed "Royal Purple" red juvenile leaves and normal leaves two weeks (May 5th to May 19th) after severance; (b) G/R values of leaf tip and base for severed "Royal Purple" mature green leaves two months (May 19th to Jul. 17th) after severance; (c) difference (Aug. 2013) of leaf stomata conductance of smoke trees between leaf tip and base areas that were separated by severing position and measured one hour after major vein being severed; (d) image temperature at tip (TS) and base (BS) areas of leaves severed at their major vein ($Sn=5$) and at tip (TC) and base (BC) areas in non-severed normal leaves ($Sn=5$) as well as image temperatures of the tip (TI) and base (BI) areas in leaves ($Sn=5$) taken within one hour after the major vein was severed at sites near Shandong University; (e) RGB image of a severed "Royal Purple" mature green leaf tested near Shandong University; (f) thermograph of the leaf shown in (e).

3.5 Leaf autumn shedding and reddening related to leaf arrangement

In our recent studies, we found that smoke trees gradually changed leaf color in autumn beginning with the bottom branches and progressing to the top branches with the same order of leaf development. Leaves on the same crown of smoke trees even showed green, light green, red and yellow colors in the period of leaf abscission. This phenomenon represents not only the spatial variance but also the structural heterogeneity of leaves. Using the RGB image classification, the G/L values for red, yellow, green and light green leaves were found to be significantly different from each other ($p=5.45E-36$). Meanwhile, G/L values were also found to decrease with color in the order from green, light green, yellow and red (Fig. 5a).

According to our leaf vein severing results mentioned herein, local water imbalance is capable of inducing leaf reddening in smoke trees. A lingering question, however, is why leaves that belong to the same tree show such different color changing patterns. In autumn, most deciduous tree species we studied began to shed leaves from the basal part of their branches/twigs, and discoloration usually occurred just before leaf abscission. We measured snap strength in the leaves of four specific color categories to analyze the relation between leaf abscission and reddening and found that these values were significantly different from each other ($p=5.82E-10$) (Fig. 5b). Snap strength also decreased with color in the following arrangement: green, light green, yellow and red. As red and yellow leaves approached the time to shed, their snap strength weakened, while green and light green leaves maintained high snap strength. Furthermore, under direct sunlight, green leaves maintained lower temperatures compared with red senescent leaves (Fig. 5c). This type of temperature difference according to leaf arrangement can be traced to summer measuring with thermography (Fig. 5d). This type of water and energy imbalance is the key component to the process of leaf reddening of the smoke tree.

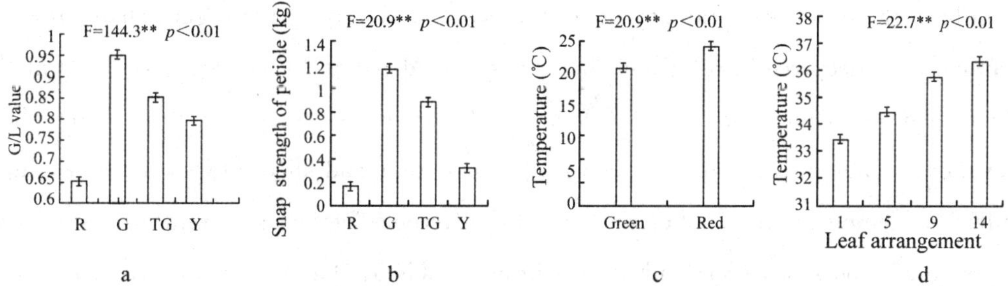

Fig. 5 Leaf feature during autumn shedding and image temperature according to leaf arrangement: (a) G/L values for red (R), green (G), light green (TG) and yellow (Y) leaves (*Sn*=20) from RGB analysis; (b) snap strength of leaf petiole for the same leaves in Fig. 5a; (c) image temperatures of green and red leaves before their shedding in the middle of Oct.; (d) image temperature of smoke tree leaves according to leaf arrangement measured in early Aug. by using thermography.

In this study, we found that in addition to smoke trees, leaf anthocyanin levels were positively correlated with leaf water loss rate in a specific time interval in 15 other tree species (8 deciduous, except the smoke tree and 7 evergreens) (Fig. 6a) ($r=0.74$) with statistical significance ($p<0.01$). Deciduous trees usually have higher levels of anthocyanin (Fig. 6a, blank circles) than evergreen tree species (Fig. 6a, black circles) ($p<0.01$). Specifically, when anthocyanin levels in evergreens were normalized to the content in deciduous trees, evergreens were found to contain only 60% of that found in deciduous leaves. Therefore, it is no surprise that evergreen species usually maintain green leaves and, hence, contain less anthocyanin during the growing season. In contrast, leaves of many deciduous tree species contain relatively high levels of anthocyanin (Fig. 6a) in Jinan, China.

With respect to the smoke tree, we found clear evidence to support the notion of a relationship between leaf reddening and water imbalance. The results from the RGB analyses for images taken in Jinan City on Nov. 11st and 12nd showed that leaf G/L values were smaller at the dry hill site (Fig. 6b-Yanzishan) compared with the moister flat site (Fig. 6b-Yinmaquan), and this difference was statistically significant ($p<0.01$) (Fig. 6b). In addition, by measuring the anthocyanin content in the saplings of many deciduous tree species planted at both the Yanzishan and the Yinmaquan sites, we found that the leaves at the branch bases had higher levels of anthocyanin compared with those from the tops of the same branch during leaf abscission, especially for the smoke tree. Similar tendencies appeared between leaf anthocyanin and chlorophyll in smoke trees from both the Yanzishan and Yinmaquan sites, respectively. At the Yanzishan site, leaf laminas were deep red in color and the leaves turned red and shed early, whereas at the Yinmaquan site, where soil was fertile and had more moisture, leaf laminas were yellow to yellow-orange in color and the leaves changed color and shed late. Another example of leaf reddening in the smoke tree was the difference between current-year sprouting twigs on old rootstock (Sp) and normal branches (Nr). The G/L values from the RGB image analyses demonstrated a statistically significant difference ($p<0.01$) (Fig. 6c). The leaves on these current-year sprouting twigs were clearly prioritizing water and nutrient allocation to maintain their deep green color. In particular, some of them maintained green leaves until winter and their leaves then wilted on green twigs. Both adult trees and recently planted seedlings of the smoke tree exhibited the same phenomenon, that is, leaves reddened from the basal to apical parts of their branches. Smaller $OD_{525/645}$ ratios were measured for tip leaves (Fig. 6d, YTT, YST and MST) and larger ratios for basal leaves. Color changes in leaves varied spatially within single leaves as well as between new and old branches in a heterogeneous fashion.

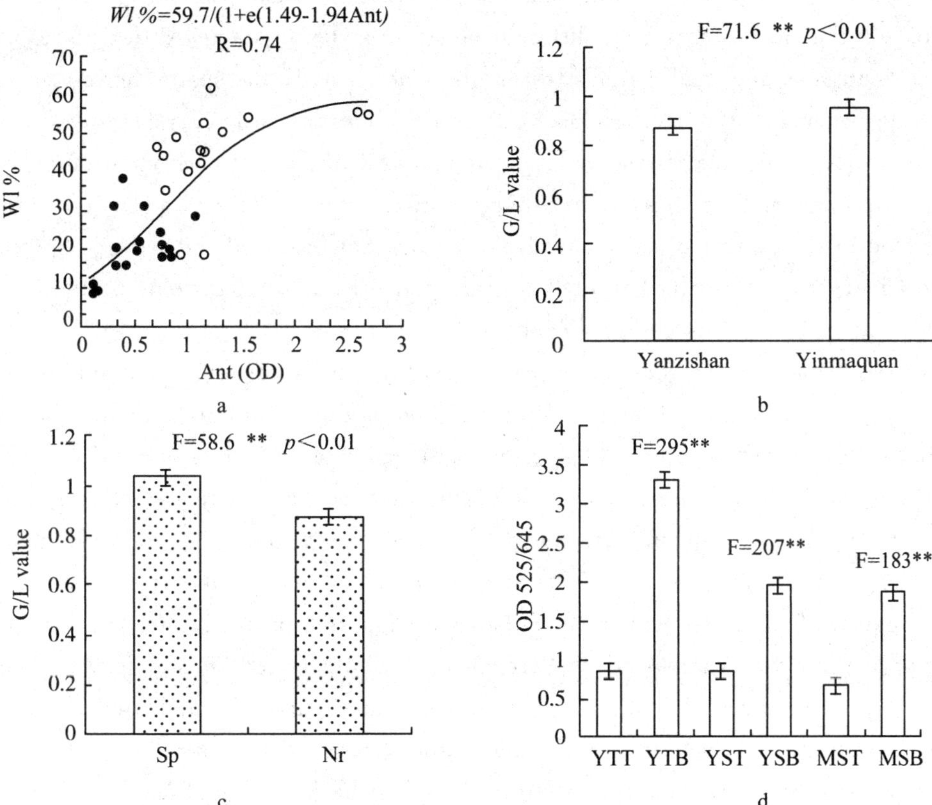

Fig. 6 Results of anthocyanin measurement and RGB analyses: (a) relationship between nine hour's percent water loss (Wl%) from detached leaves of 15 deciduous (blank)/evergreen (black) tree species and anthocyanin content expressed in optical density values (Ant(OD)); (b) G/L values for leaves from smoke trees planted at a dry hill site (Yanzishan, *Sn* = 10) and a moisture flat site (Yinmaquan, *Sn*=15); (c) G/L values for leaves from current year sprouting shoots (Sp, *Sn*=6) and normal smoke trees (Nr, *Sn*=6); (d) comparison between anthocyanin and chlorophyll in leaves of smoke trees planted at different site conditions, evaluated using the $OD_{525/645}$ value in Oct. during the process of leaf shedding, whereas YTT and YTB are the tip and base leaves on adult tree at Yanzishan, YST and YSB are the tip and base leaves on current year planted seedlings at Yanzishan, and MST and MSB represent the tip and base leaves of current year planted seedlings at Yinmaquan nursery (*Sn*=3).

4 Discussion

Leaf coloration is affected by genetic components and the surrounding environment, which includes a variety of factors such as light irradiation, temperature, transpiration cooling (Jones and Leinonen, 2003; Grant *et al.*, 2006), energy metabolism (Wang and

Yamamoto, 2010) and so on. The venation structure is essential to the spatial distribution of water as leaves that have different venation structures show different responses to local vein severance and water stresses. In other words, the spatial heterogeneity of sites experiencing water stress induces an uneven allocation of energy, temperature and leaf coloration. Thus, a sparse venation structure without network connection makes smoke trees particularly sensitive to vein severance and water stress, a sensitivity that is illustrated in the pattern of color changes in leaves that begins at the leaf tip, spreads to the leaf margins and finally enters the interstitial area, a similar color pattern change that was described 100 years ago (Yapp, 1912).

Different patterns of vein severance induce different water stress and energy imbalances on partial areas of leaf lamina. Water and energy gradients cause the leaf lamina to exhibit different colors, which range from light green to red. The pairing of advanced reddening and high temperature areas indicates that leaf lamina reddening is related to water and energy metabolism. After vein severance, the high consistency with which we observe a relationship between a water stressed area and persistent/advanced leaf reddening suggests that a persistent water imbalance triggers a protective response in partial areas of leaf lamina. An inspection of transverse sections of red juvenile leaves of smoke trees, staghorn sumacs and purple blow maples under a microscope show that the red layer is formed at the epidermis or palisade tissue to reduce levels of red light absorbance (Wang *et al.*, 2014). In addition, we found that the back surface of leaves in the smoke tree and staghorn sumac redden when they are turned inwards for several days. In the process, anthocyanin in the red epidermis and surface tissues sustain the leaves to release the stress caused by an overload of light energy (Hoch *et al.*, 2001; Feild *et al.*, 2001; Hughes *et al.*, 2007) and water stress (Chalker-Scott, 2002). An imperfect cuticle layer, rapid water loss and sparse stoma in juvenile leaves are the trigger factors of anthocyanin accumulation in the epidermis and surface tissues.

The leaf shedding process involves the formation of an abscission layer and results in vascular obstruction, which is similar to leaf severing at the petiole base. Smoke trees shed their leaves soon after the leaves turn red, similar to many other tree species. Leaf color change in deciduous tree species in autumn is considered to be the result of a shortening of the photoperiod and a decrease in temperature (Howe *et al.*, 1995; Pietrini and Massacci, 1998; Close *et al.*, 2002). However, the old senescent leaves of many evergreen tree species usually redden and shed concentratedly in late spring and early summer (Kozlowski, 1973) when the photoperiod is lengthening and the temperatures are rising. These shedding leaves are mostly located at the base of the branches where there is less light irradiation. According to this study, this phenomenon is easily understood as a result of a large amount of water being transpired from juvenile leaves. As older leaves are unable to effectively compete for water, they are shed to maintain balances of both water and energy throughout the entire tree. During the shedding process, leaves face an obstruction of water transport, which leads to the onset of the protective process (Gould *et al.*, 1995; Feild *et al.*, 2001; Hughes *et al.*, 2007). In addition, the juvenile leaves of many tree species often redden regardless of the season.

Some vigorous leaves with deep green color, particularly those on current-year sprouting twigs, wilted while they were still attached to the twig and did not begin abscission until winter. This phenomenon has been described for deciduous trees and is known as green withering. Trees that show green withering are often subject to severe cold waves as well as frost days in late autumn. As these conditions may be similar to those experienced by the smoke trees in this study, we inferred that leaf reddening is usually consistent with the formation of the abscission layer.

The early and widespread reddening of ornamental tree species, such as smoke trees, planted at dry and barren mountain locations also implies that leaf reddening is related to the water and energy imbalances. Under water and energy imbalance conditions, red leaves often persistently maintain red coloration while green mature leaves show advanced reddening as a protective requirement against excess light or heat energy (Hoch *et al.*, 2001).

In brief, leaf reddening in smoke trees is related to water and energy imbalances (Chalker-Scott, 2002). An overload of water consumption triggers a failure in the transpiration system such that leaves are not effectively cooled. This, in turn, leads to an energy imbalance that induces a photoprotective response and an accumulation of anthocyanin in sensitive areas with a different color than normal leaves. Therefore, the mechanism of leaf reddening in smoke trees may be a photoprotective response to water and energy imbalances induced by extra-transpiration, vascular obstruction and imperfect leaf structure.

References

[1] Archetti M. The origin of autumn colors by coevolution[J]. Journal of Theoretical Biology, 2000, 205: 625-630.

[2] Chaerle L., Van-Der-Straeten D. Imaging techniques and the early detection of plant stress[J]. Trends of Plant Science, 2000, 5(11): 495-501.

[3] Chalker-Scott L. Environmental significance of anthocyanins in plant stress responses[J]. Photochemistry and Photobiology, 1999, 70: 1-9.

[4] Chalker-Scott, L. Do anthocyanins function as osmoregulators in leaf tissues[J]? Advances in Botanical Research, 2002, 37: 103-127.

[5] Close C. L., Holz G. K., Brown P. H. Effect of shadecloth tree shelters on cold-induced photo inhibition, foliar anthocyanin and growth of Eucalyptus globulus Labill and E. nitens (Deane and Maiden) maiden seedlings during establishment[J]. Australian Journal of Botany, 2002, 50: 15-20.

[6] Field T. S., Lee D. W., Holbrook N. M. Why leaves turn red in autumn: The role of anthocyanins in se-

nescing leaves of red-osier dogwood[J]. Plant Physiology, 2001, 127: 566-574.

[7] Grant O. M., Chaves M. M., Jones H. G. Optimizing thermal imaging as a technique for detecting stomatal closure induced by drought stress under greenhouse conditions[J]. Physiologia Plantarum, 2006, 127: 507-518.

[8] Grant O. M., Tronina L., Jones H. G., Chaves M. M. Exploring thermal imaging variables for the detection of stress responses in grapevine under different irrigation regimes[J]. Journal of Experimental Botany, 2007, 58(4): 815-825.

[9] Hoch W. A., Zeldin E. L., McCown B. H. Physiological significance of anthocyanins during autumnal leaf senescence[J]. Tree Physiology, 2001, 21: 1-8.

[10] Howe G. T., Hackett W. P., Furnier G. R., Klevorn R. E. Photoperiodic responses of a northern and southern ecotype of black cottonwood[J]. Physiologia Plantarum, 1995, 93: 695-708.

[11] Hughes N. M., Morley C. B., Smith W. K. Coordination of anthocyanin decline and photosynthetic maturation in juvenile leaves of three deciduous tree species[J]. New Phytologist, 2007, 175: 675-685.

[12] Jones H. G., Leinonen L. Thermo imaging for the study of plants water relation[J]. Journal of Agricultural Meteorology, 2003, 59(3): 205-217.

[13] Kozlowski T. T. Shedding of Plant Parts[M]. New York: Academic Press, 1973: 1-117.

[14] Kytridis V-P., Manetas Y. Mesophyll versus epidermal anthocyanins as potential in vivo antioxidants: Evidence linking the putative antioxidant role to the proximity of oxy-radical source[J]. Journal of Experimental Botany, 2006, 57(10): 2203-2210.

[15] Pietrini F., Massacci A. Leaf anthocyanin content changes in Zea mays L. grown at low temperature: Significance for the relationship between quantum yield of PSⅡ and the apparent quantum yield of CO_2 assimilation[J]. Photosynthesis Research, 1998, 58: 213-219.

[16] Prytz G., Futsaether C. M., Johnsson A. Thermography studies of the spatial and temporal variability in stomatal conductance of *Avena* leaves during stable and oscillatory transpiration[J]. New Phytologist, 2003, 158: 249-258.

[17] Slavik B. Methods of Studying Plant Water Relations[M]. Berlin, Heidelberg, New York: Springer-Verlag, 1974: 284-285.

[18] Wang F., Yamamoto H., Ibaraki Y., Iwaya K., Takayama N. Evaluation Ginkgo leaf necrosis and asymmetric crown discoloration induced by Typhoon 0613 with RGB image analysis[J]. Journal of Agricultural Meteorology, 2009, 65(1): 27-37.

[19] Wang F., Yamamoto H. Detecting leaf and twig temperature of some trees by using thermography[J]. Spectroscopy and Spectral Analysis, 2010, 30(4): 2914-2918. (In Chinese)

[20] Wang F. Persistent and advanced reddening of sweetgum leaves after major veins severing[J]. Journal of Forestry Research, 2010, 21(4): 465-468.

[21] Wang F. Transpiration cooling failure and discoloration of major vein severed sweetgum leaf[J]. Journal of Beijing Forestry University, 2013, 35(1): 72-76. (In Chinese)

[22] Wang F., Yamamoto H., Li X. M., Zhang J. Q. Leaf reddening of sweetgum in water imbalance[J]. Ecological Informatics, 2014, 19(2014): 47-51.

[23] Yapp R. H. *Spiraea Ulmaria L.* and Its Bearing on the Problem of Xeromorphy in Marsh Plants[J]. Annals of Botany, 1912: 815-870.

（原文发表于 *Urban Forestry & Urban Greening*, 2015, 14: 80-88）

第四部分　叶尖叶缘枯萎

Transpiration surface reduction of Kousa dogwood trees during seriously losing water balance

WANG Fei[1], Haruhiko Yamamoto[2]

(1. The United Graduate School of Agriculture Science, Tottori University;
2. Faulty of Agriculture, Yamaguchi University)

1 Introduction

Plants usually live in the contradictory processes in receipting carbon or energy resources and water loss. To maintain a higher photosynthesis and carbohydrate production per land area, it needs additional leaf areas, which implies more water and nutrition consumption. Most of water absorbed from soil is lost by plant transpiration and less 5% is used in metabolism and growth. Therefore, transpiration has been ever regarded as an unavoidable evil since it causes water deficits and injury by dehydration (Kramer, 1983). It is also considered beneficial because it acts as transpiration cooler to avoid leaf temperature to over raise, causes the ascent of sap, and increases the absorption of minerals (Clements, 1934). Plant tissues dissipate heat by three main processes—emission of long-wave radiation, convection of heat and transpiration of water, of which transpiration tends to be the most effective process of heat dissipation of plant tissues, particularly at the midday. High plant temperatures (>40 ℃) are almost invariably associated with the cessation of transpiring cooling, following stomata closure in response to drought (Fitter *et al.*, 2002). Therefore, the transpiration cooler fail during the serious drought stress seems lethal to plants. In addition, increase in temperature alone tends to cause an increase in the rate of transpiration through its effect on saturation water vapor density (Fitter *et al.*, 2002). Under these kinds of conditions, excessive leaf area usually causes losing balance of energy and water so as to be dangerous to the plants' lives. A lot of plant species respond to the unfavorably extreme hot and droughty stress by transpiring surface reduction (TSR) to maintain the water balance of left parts of them. TSR has been considered as a hydroecological factor for a long time (Orshan, 1954). It is also thought as an approach of reducing radiation acceptation to maintain the energy balance of plants (Kozlowski, 1973). It can be seen in various patterns, leaf or branch shedding for many deciduous trees, even evergreens (Addicott, 1982; Rust, 2004) and the death of aboveground for most annuals and grasses etc. (Kozlowski, 1973; Bhat, 1986). Some tree species respond to the unfavorably extreme droughty environment with partial leaf scorch (Gunthardt-Goerg *et al.*, 2007; Vollenweider *et al.*, 2006) as the special leaf structures and adaptive mechanism. This kind of response can be remarked as a response of partial aboveground death, or a grass like response. The homogeneity of these kinds of adaptation was shrinking back of the living parts from distal to proximal to respond to the water deficit and excessive radiation acceptance. Through partially withering leaves the plants reach the trade-off among water loss, radiation acceptance and CO_2 flux to survive from the extreme hot and droughty environment.

Based on the conventional explanation for dying of extreme drought stresses, plants usually lose turgor during the stresses at first. As plants become severely dehydrated they are less likely to regain turgidity during the night, often resulting in permanent wilting of leaves (Kozlowski, 1972). Tension to the vascular tissue and parenchyma increases as drying progresses, eventually disrupting water flow; vascular and mesophyll cells collapse and tissues wither, collapse and die (Treshow, 1970). But unlike the plants that respond to drought stress as a whole, or tissues and organs evenly changed, many of them show uneven responses from distal to proximal. Gradually drying back of living tissues during unfavorably extreme drought stresses takes place accompanying with the occurrence of clear defense barrier. It results in partially shearing the soil-plant-atmosphere-continuum field (SPACF), which perfects the process of water absorption, transportation and transpiration from soil via plant body into atmosphere, of these plants. Many landscape trees including bamboos (*Sasa sp.*) and dogwoods (*Cornus sp.*), even some succulent plant species (Addicott, 1982) usually appear this kind of symptom.

Following the abnormal droughty spring, hot and dry summer in 2007, some landscape trees showed abnormal status, especially the trees planted on the limited site, such as coarse sand soil, rocky site, wall-flower bed and the site with root growing limitation etc. Many Kousa dogwood trees appeared gradual leaf scorch from distal to proximal. During the spring and summer in 2008, leaves' gradual scorch-back was also observed on some transplanting shocked and previously planted dogwood trees in Yamaguchi.

The heterogeneousness of leaf scorch-back and leaf color varied from tip to base or from distal to proximal, which caused them difficult to be directly measured. The flexibility of RGB image analysis made itself suitable to measure the leaves differentially. In the study, the leaf images were equally divided into ten sections from proximal to distal and the "switch-off" type of threshold responsive functions (Thornley, 1976) was established to describe the gradually scorched leaves. The water contents were also analyzed by similar differential pattern. By using the image pixel analysis and water content measurement of scorched leaves, the TSR process of the dogwood trees during the abnormal extreme drought event has been described quantitatively.

2 Materials and methods

Dogwood trees, about 7 years old with height of 3－4 meters, were observed to study the leaf scorching response to the extreme drought stresses in 2007 and 2008. Some newly planted dogwood saplings were also observed to research the transplanting shock during the spring and early summer in 2008. They are all planted around a park, which is the ancient riverbed (Sakaue *et al.*, 1972; Miura *et al.*, 1972) and the former athletic track in Yamaguchi city, Japan. Branch dieback occurred on many landscape tree species planted around including the dogwood trees. On the stems of some dogwood trees, the trace of scale insect parasite was also found. It suggested that improper site condition made the trees sensitive to environmental changes.

The RGB analysis of leaf scorch-back for these dogwood trees was based on the leaf images obtained from scanning with a scanner (Canon D125u2). Ten leaves were typically sampled from each tree and 60 stocks were sampled in total. The leaf reduction area percentage (LRAP) was determined by image pixel method. It is a proportion of scorched area to overall leaf area and measured by getting pixels of overall leaf and the green part for each leaf with Photoshop. The LRAP was calculated by Equation (1):

$$\text{LRAP} = 100 - \left(\frac{\text{pixels for green area of leaf}}{\text{pixels for overall leaf}} \times 100\right) \tag{1}$$

To analyze the characteristics of leaf scorch, the same leaf images above mentioned were used to measure the G/R value, an index in RGB image analysis (Adamsen *et al.*, 1999). Before getting RGB pixel data, the image was hand prepared by eraser of Photoshop to remove the background and objects except the objective leaf. Then leaf images were equally divided into 10 sections from base to tip. The Red (R) and Green (G) values for each section were read from the average histogram value of Photoshop. The G/R value was calculated by Equation (2):

$$\text{G/R} = \frac{\sum_{i=0}^{255} N_i \times i / \sum_{i=0}^{255} N_i}{\sum_{j=0}^{255} N_j \times j / \sum_{j=0}^{255} N_j} \tag{2}$$

where, N_i is the pixel number in i (green) gradation, $i=0, 1, 2, \ldots, 255$. N_j is the pixel number in j (red) gradation, $j=0, 1, 2, \ldots, 255$.

By repeated regression test, the relative G/R (RGR) decreased nonlinearly from proximal to distal and could be modeled by logistic threshold responsive Equation (3) for scorched leaves.

$$\mathrm{RGR}(n) = \frac{k}{1+e^{a-rn}} \tag{3}$$

where, RGR [Equation (4)] stands for relative G/R. RGR(n) is the RGR value at n section. Obtained by regression process, r and a are constants and k is the maximum value that the RGR can reach. n ($n=1, 2, \ldots, 10$) is the number of leaf sections.

$$\mathrm{RGR}_i = \frac{100 \times (\mathrm{G/R}_i - \mathrm{G/R}_{\min})}{(\mathrm{G/R}_{\max} - \mathrm{G/R}_{\min})} \tag{4}$$

in which, $\mathrm{G/R}_i$ is the G/R value for i section, $\mathrm{G/R}_{\min}$ is the minimum G/R value of all sections and $\mathrm{G/R}_{\max}$ is the maximum G/R value of all sections.

In addition, two defense barriers appeared on some leaves (Fig. 1d) and two responsive equation lines were established. For the second one, the responsive equation was made by direct measurement and regression. For the first one, the image of the leaf was firstly restored to the first scorched status by filling the scorched area between the first and second barrier with average green value of the leaf.

Leaf water relation was also researched by measuring water content before and after scorch-back. Small twigs were cut from selected trees and then taken back to Lab with vinyl-bags. The water content for single leaves or leaf sections from newly transplanted saplings and normal growing trees were measured by rapid weighing method with 1/10000 g electronic balance in room. The weight of sampled leaves or leaf sections were weighed after sampling from field site without delay. After obtaining the dried weight of them, the water content was calculated by Equation (5):

$$\mathrm{WC}\% = \frac{\mathrm{FW}-\mathrm{DW}}{\mathrm{FW}} \times 100\% \tag{5}$$

where, FW is the fresh weight of sampled leaf and DW is the dried weight of the same leaf.

Leave sat threshold status of leaf scorch-back or after scorch-back were hoof-shapely cut into seven or eight (according to the leaf size) sections from proximal to distal during the persistent dry and hot summer days in August in 2008. The water content for each section was calculated by Equation (5) to study the variant tendency of water content from distal to proximal of the leaves.

Five functional and water saturated mature leaves for each tree and shrub of Kumazasa bamboo (*Sasa Veitchii Carr.*), Kousa dogwood, sweetgum (*Liquidambar styraciflua L.*), Japanese blue oak (*Quercus glauca Thunb.*) and sasanqua were picked up from normal growing trees at a rainy day to study the water loss process of isolated leaves. They were dehydrated under the indoor environment of RH 60%−70% and AT 25−30 ℃ and weighed in the planned time interval. Water loss percentage was also calculated by Equation (5). By this method, the leaf cuticle transpiration characteristics of these five tree species were estimated and observed. Also every five normal Kousa dogwood leaves were orderly cut into ten sections from proximal to distal after losing water 12.3%, 29.5% and 45.3% respectively. The water content for each section was calculated by Equation (5) to study the evenness of water loss in leaves.

In order to find the relation between TSR and the extreme drought event, the daily meteorological data during 2007 and 2008 for Yamaguchi observatory, 1.2 km from the investigated trees, were obtained from Automated Meteorological Data Acquisition System of Japan, and the first ten records of maximum or minimum value from 1976 to 2007 as well. The aridity index of eleven days (AD11) was calculated by Equation (6) for the meteorological data from April 1st to August 31st, in both 2007 and 2008 respectively.

$$AD11_i = \sum_{j=0}^{10} MT_{i+j} / \sum_{j=0}^{10} PR_{i+j} \tag{6}$$

where, $i=1, 2, \ldots, 153$ and $i=1$ at the April first. MT is daily maximum temperature and PR is daily precipitation.

3 Results and discussion

3.1 TSR of Kousa dogwood after persistent droughty weather in 2007

Kousa dogwood is a well-known landscape tree species with showy flowers and widely cultivated along street and around house in Yamaguchi. But it may appear leaf scorch when affected by drought, after root injury and transplanting shock etc. The similar symptoms appeared on many dogwood trees after the effect by dry and hot summer in 2007 in Yamaguchi City. Tip and/or edge leaf scorch appeared on many of dogwood trees, which led to their crown's discoloration in different scales.

It was observed that the responses of the dogwood varied significantly from trees to trees and among leaves (Fig. 1a). The threshold responsive equation for image RGR value of their leaves with different scorched areas (Fig. 1a Leaf 2, Leaf 3 and Leaf 4) presented different inverse logistic curves (Fig. 1b. Leaf 2, Leaf 3, Leaf 4). It indicate that the injury did not evenly distribute on the leaves and the scorched area bound from distal to proximal, which is the typical scorch-back character. Carefully observing the scorched leaf, we found that apparent defense barrier existed on the leaf surface and the barrier lines also arranged from distal to proximal catastrophically (Fig. 1a). The shape of responsive function varied from inverse sigmoid shape to rectangular hyperbola shape (Fig. 1b) as the scorch became severe. Meanwhile, the leaf area was reduced through scorching the part outside the barrier. According to the color analysis of scorched part, only one major defense barrier could be observed on most of leaves (Fig. 1c). For seriously injured leaves, two (Fig. 1d) even three or more could be seen. It indicated that the defense barrier withdraw back from distal to proximal gradually until successfully controlling the scorching and leaving a series of unsuccessful defense traces (Fig. 1a, Leaf 3, Leaf 4). The threshold responsive curves for both non-scorched leaf and entirely scorched leaf (Fig. 1a Leaf 1, Leaf 5) appeared a tendency of straight lines slightly slanted and laid on top and bottom of the coordinate separately (Fig. 1b). It should be the characteristics of the trees with no-surface area reduction and with surface area thorough reduction separately.

Among the leaves that appeared multi barrier line or defense trace, most of them contained two belts differently colored and separated by two defense barriers (Fig. 1d). It indicated two scorched periods occurred from the sprouting of the leaves and showed different responsive function curves.

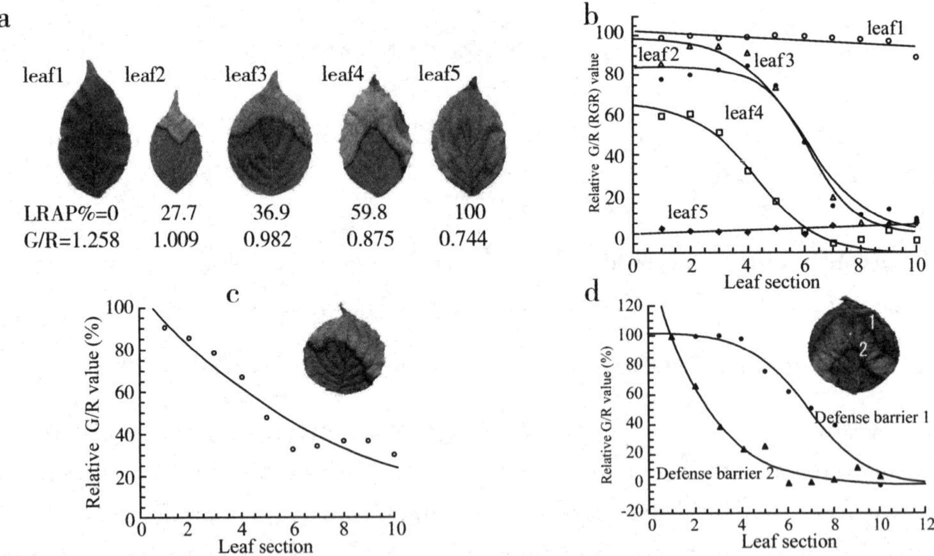

Fig. 1 Variant characteristics of Kousa dogwood leaf scorch from proximal to distal and defense barriers: 1a presents five leaves from different trees with different leaf reduction area percentages (LRAP), and LRAP are 0, 27.7, 36.9, 59.8 and 100 respectively. Image Green/Red (G/R) value ranges from maximum 1.258 to minimum 0.744. 1b shows the responsive curves of relative G/R (RGR) for these leaves, with the characteristics of typical logistic curves for scorched leaves (Leaf 2, Leaf 3 and Leaf 4), and direct lines for overall green and entire brown leaves (Leaf 1 and Leaf 5). 1c shows a leaf with one defense barrier and its RGR thresh-

old responsive curve. 1d presents a leaf with two barriers constructed in May and August 2007, and related threshold responsive curves of RGR value

Despite the fact that the characteristics of the defense barrier and leaf scorched area varied significantly, the total living area of leaves commonly reduced. Calculating the image LRAP, it presented 2.7% and 38.6% for the first and second scorched phase for all of the sampled trees. In total, the reduction percentage of leaf area was about 40% of total sampled leaves. Even if it was coincidence that the precipitation during the first nine months was about 40% less than that of normal years, the relevance between the leaf scorch of Kousa dogwood trees and less precipitation should be less doubt. It was clear that Kousa dogwood trees manifested grass-like response to it and showed serious leaf scorch-back during sudden dry and hot environment under the insufficient water supply. It was evident that the leaf scorch became serious as the stresses increased and resulted in the decreasing of the total leaf areas of the dogwood tree. It indirectly decreased the water or precipitation requirement and made the living parts of entire tree receive less radiant energy. It seemed the green parts of scorched leaves maintained active status and as the environment became favorable they restored vigorous immediately. It was observed that the green part of some scorched leaves of Japanese blue oak hit by Typhoon 0613 maintained normal function even after two years in the same city.

3.2 TSR of some Kousa dogwoods during transplanting shock in 2008

After transplantation, the successful survival of trees mostly depends on rapidly establishing the perfect root system. If not, the new sprout leaves may suddenly dry out or scorch-back under sudden drought environment for the reason of losing water balance. Sufficient precipitation, 116% of the normal, during the first half year in 2008 made the dogwood trees appear different responses from that in 2007. Almost no leaf scorch-back symptoms occurred on the same dogwood trees observed in 2007 before summer days in 2008 (Fig. 2a). Only some newly complementarily planted trees showed the gradual leaf scorch-back symptoms during the sudden increasing of the temperature and no rain weather on May 8th (Figs. 2b, 2c). Fig. 2 shows different dogwood leaves from the trees transplanted in winter in 2007 and their responsive lines of RGR value during the spring in 2008.

It was observed that the leaves from normal growth trees appeared a level responsive curve of RGR value in Fig. 2a, while under the stress of transplanting shock the leaves desiccated from tip to base and the appearance became uneven from distal to proximal (Fig. 2b). A black shade layer between dried and non-dried area was observed and their responsive curves slanted at tail end. In this situation, although the leaf tip had dried out, the color of it still remained green. It seemed that water loss was too fast to change the chlorophyll. Three days later, the leaf tips became deep gray and a typical RGR responsive function of inverse logistical curve or scorch-back symptom emerged (Fig. 2c). During the shock, a lot of seriously hit leaves dried out after several days' persisting warmer and no rain weather at the beginning of May. Soon after, the coming of the Japanese rainy season and about 350 mm monthly precipitation in June promoted the new sprouting of small leaflets with long and narrow tips on the tree. It was calculated by image pixel method that the leaf area of the transplanting shocked tree was only 38.6% of that before the shock. After the end of the Japanese rainy season in the beginning of July and about ten days' persistent drought and hot weather, the remained leaves and small new sprouting leaves scorched back once again (Fig. 3a). Some of them also presented two defense barriers on leaflet (Fig. 3b) after two periods of shock. The RGR responsive lines showed a similar tendency as the first shock during May.

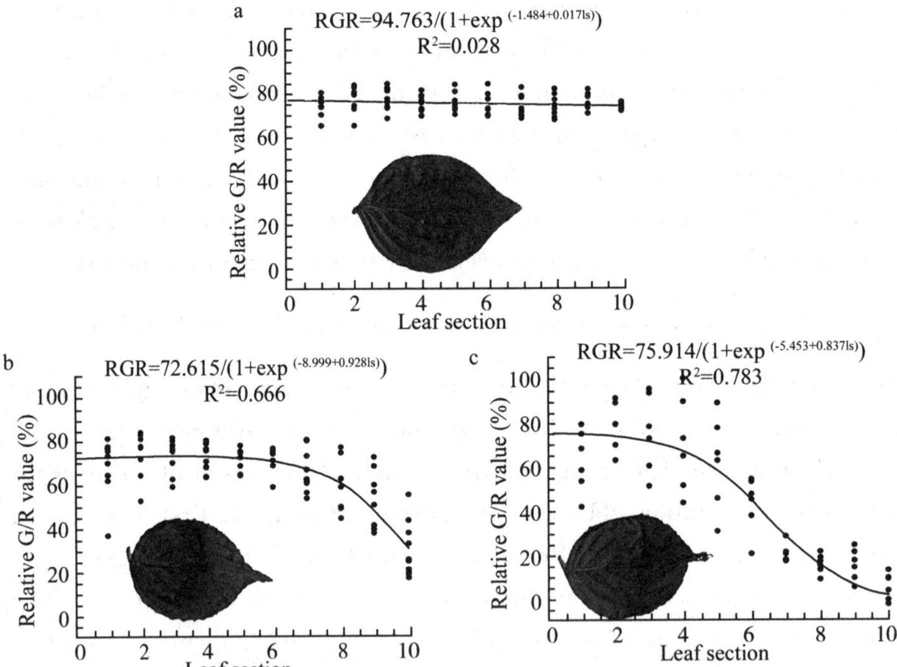

Fig. 2 Threshold responsive curves of relative G/R (RGR) value for leaves from normal growth (2a) and newly transplanted Kousa dogwood trees (2b, 2c): 2b represents a threshold responsive curve of leaves during the sudden aridity increment stress in May 8th 2008, and 2c shows the threshold responsive curve of transplanting shocked leaves on the same tree three days later.

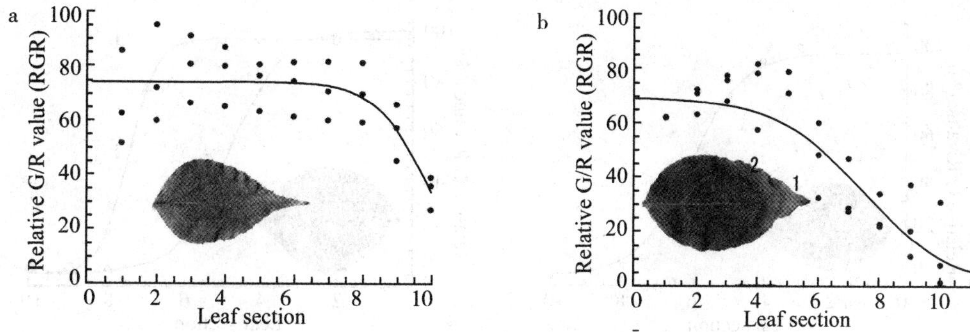

Fig. 3 Transplant shocked characteristics of the new sprouting leaf after Japanese rainy season in 2008 (3a), the leaf with two defense barriers (3b) and the threshold responsive curve of relative G/R (RGR) values for them (3a, 3b).

However, the leaves with scorched area in both Figures 1a, 1c and Figures 2c, 3a, 3b appeared a significant difference by contrast. In Figures 1a and 1c, almost all of the scorched leaves showed brown or light brown scorched part, while in Figure 2 and Figure 3 the scorched part of the leaves presented gray or dark gray color. The former occurred in extreme droughty and hot August in summer and the latter emerged in a humid and warm May in spring or after the Japanese rainy season. In Figure 2c, the defense line can't be seen clearly as the ones in Figures 1a and 1c. However, it seemed to evade the lethal drought by scorching partial of leaves from distal toward proximal and all of them acted as a result of surface area reduction to survive from seriously losing water balance.

3.3 TSR of Kousa dogwood after persistent hot and droughty weather in 2008

Then one month's persistent no rain and high temperature from July 14th to August 14th, the hottest days in a year, not only induced the transplanted dogwood trees to reduce their transpiring surface area again, but also led to partial of the previously planted dogwood trees' leaf scorch-back from distal to proximal (Figs. 4a, 4b). The symptoms of leaf scorch-back appeared significantly similar to that occurring in the summer of 2007 (Figs. 4a, 4b, 1c, 1a), which also showed light brown scorched leaf areas. It was observed that the defense barrier almost simultaneously appeared on leaves during one night, although the severity of symptoms differed from leaf to leaf. The dogwood trees that showed symptoms of leaf scorch-back in the middle of August in 2008 were less than 1/2 of that in 2007. The total leaf area reduction in 2008, only 13.2% of the entire leaves, was less than 1/3 of that in 2007, although more trees appeared angular leaf spot disease in 2008.

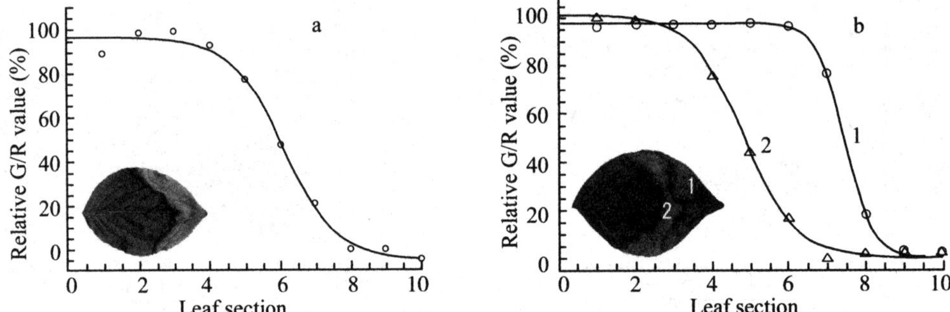

Fig. 4 Leaf scorch-back characteristics of previously planted dogwood trees during the droughty and hot summer from July 14th to August 14th, 2008 and the threshold responsive curves of relative G/R value for them. Some of them appeared one defense barrier (4a) with a few of unsuccessful defense barrier line constructed in one night. Some presented two defense barriers (4b) with different colored scorch belts constructed in different periods (1 on August 2 and 2 on August 12). The leaves in Fig. 4 were sampled from the same trees as Fig. 1 in a park in Yamaguchi City.

3.4 Water relation during TSR of Kousa dogwood

The premature response of Kousa dogwood to extreme hot and dry environment presented leaf gradually scorching or drying from distal to proximal (Fig. 1, Fig. 2, Fig. 3). This kind of symptom is also considered as leaf scorch-back in this paper, which looks like the dieback of branches. In fact, the serious leaf scorch in overall crown of trees and shrubs often triggers dieback of branches. Therefore, it is more proper to name the leaf scorch from distal to proximal as leaf scorch-back.

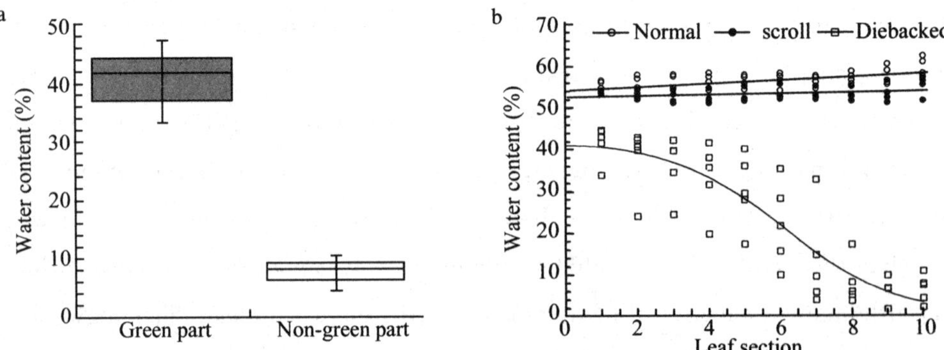

Fig. 5 Water content for the pre-scorched leaves and after-scorched leaves of Kousa dogwood: 5a presents the leaf water content for green part and non-green part of scorched leaves ($n=10$). 5b show the variation tendency of water content from proximal to distal for Kousa dogwood leaves during the transplanting shock.

Normal growth leaves (○-○) were also measured as the comparison to the scorched leaves (□-□) and scrolled leaves (●●). The SPACF was sheared by this kind of dried back (scorch-back) of the leaves.

The leaves in scorch-back were usually distinguished into green part and non-green part or the fresh part and dried part. There should be big difference of water status between them and it is proved by the measurement of water content (Fig. 5a). It seemed that there existed a great resistance between fresh part and dried part at the boundary (defense barrier) to prevent the further water, nutrient and other resources loss and intrusion of pathogens. The deep colored defense barrier, functioned as the protective layer during the normal leaf shedding, can be clearly seen from leaves scorched back in Fig. 1a, Fig. 1c, Fig. 2c, Fig. 3b, Fig. 4a, and Fig. 4b.

This kind of scorch-back generally originated from the parts of the leaf that were farthest from the main vesicular channel (Yapp, 1912). During the process of serious drought, the shrinkage of the distal tissues facilitated their separation from proximal tissue (Addicott, 1973). After persistently losing water balance, the SPACF significantly withdrew back gradually and it resulted in the SPACF shearing. It can be clearly expressed by the leaf water content decreasing from distal to proximal for the scorched leaves (Fig. 5b). By comparison, the normal leaves not only have more water content in entire leaf but also tend to get higher water content at the distal part. The water content of seriously scrolled leaves was lower than that of normal leaves especially at the tip (Fig. 5b).

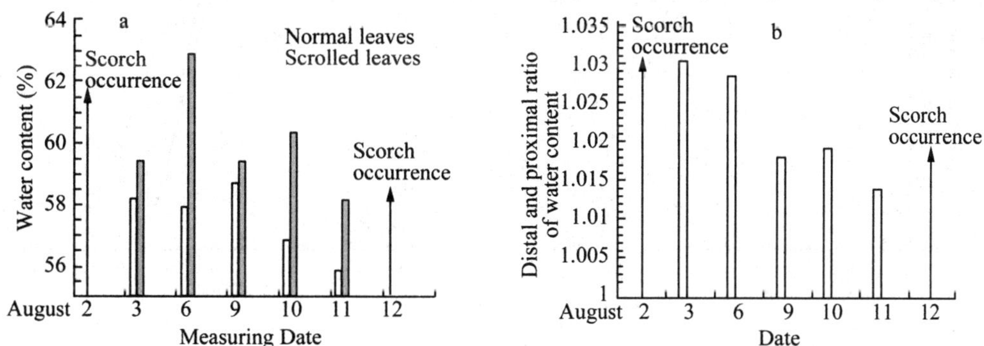

Fig. 6 The variation tendency of water content for normal growing leaves and scrolled leaves (6a) and the water content ratio between distal and proximal of previously planted Kousa dogwood trees during seriously hot and dry summer weather conditions and at the interval of two times of leaf scorch occurrence (6b). In 6b, the distal and proximal ratio was the proportion of the water content between distal and proximal parts of the leaves, which were divided into three parts—distal, middle and proximal. The data in 6b were the average value from five trees and ten leaves for each tree.

Leaf scorch-back of previously planted Kousa dogwood trees seemed to extend a prolonged process. During this process, leaves usually appeared different degrees of scrolls for a long time and maintained lower water content than the normal growing trees (Fig. 6a). Although there was a tendency of lower distal and proximal ratio of water content as the hot and droughty conditions persisted, the value of this ratio remained above 1.0 (Fig. 6b) until a few leaves showed significant leaf scorch-back on a tree during a sudden hot and dry weather (Fig. 7a). At this period of time, some of the leaves appeared evident tendency of water content variance between leaf tip and base with a regression line of water content slightly down slanted from proximal to distal (Fig. 7a). But the logistic threshold response curve (Fig. 7b) could not be seen till serious leaf scorch-back occurred on the tree. It was observed that leaf scorch-back happened only during one night and not only one defense barrier appeared on leaves, but also two even more barrier lines could be seen on the same leaf. Before the last defense barrier, even one or more unsuccessful lines were constructed on the leaf until thoroughly controlling the leaf scorch-back. It seemed that the establishing of the defense barrier and SPACF shearing was an active protection process. It suggested that under the serious tension of enlarged water gradient leaves responded by actively shearing the SPACF to protect themselves from further water loss and reduce the transpiring surface.

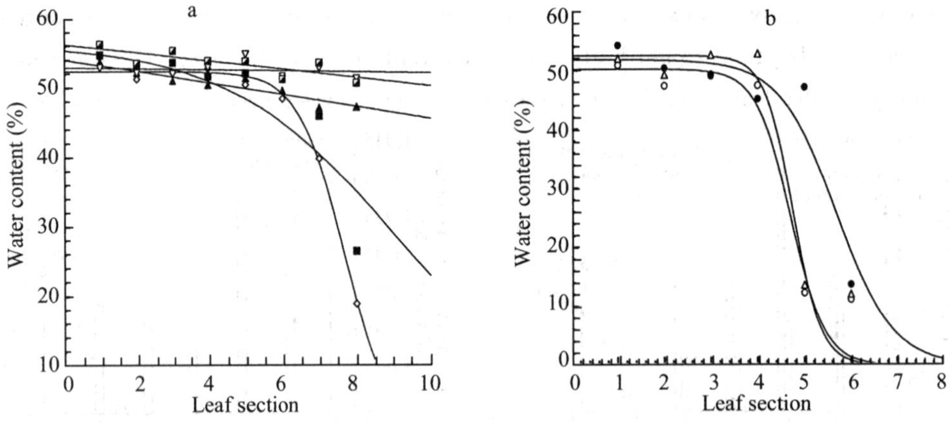

Fig. 7 The variation tendency of water content at the threshold status of leaf scorch-back during the persistent dry and hot summer days in August 2008. 7a shows the water content for the leaves before the serious scorch-back, which contained early stage of scorched leaves (◇-◇, ■■), seriously wilted leaf (▲-▲) and wilted leaves (▽-▽, ◈-◈). 7b contains the leaves with dual defense barriers (○-○, △-△) and single defense barrier (●●) after the significant scorch-back occurred as well as the significant defense barrier appeared.

The isolated leaves also showed a tendency of gradually drying from distal to proximal (Fig. 8a), although there was no significantly visible symptoms appearing on the leaf (Fig. 8b). It suggested that the coherent tension could not only lift the water to high elevation but also withdraw the water back from distal to proximal under the extreme water stress condition. Under this kind of acute water loss situation, there was no time for the leaves to respond and they had no ability to establish the defense barrier as the intact leaves. The isolated leaves, which artificially brook up the SPAC from proximal making the leaves loss water resources, dried at natural environment and showed different RGR curves (Fig. 8b) comparing to the transplanting shocked or scorched leaves (Fig. 2c). Almost a level line had been obtained being parallel to that of normal leaves (Fig. 8b). It also indicated that the defense barrier occurred only on the intact leaves and the interaction of the SPACF withdrew back and the self-defense response was the key of the TSR of Kousa dogwood during sudden drought stresses.

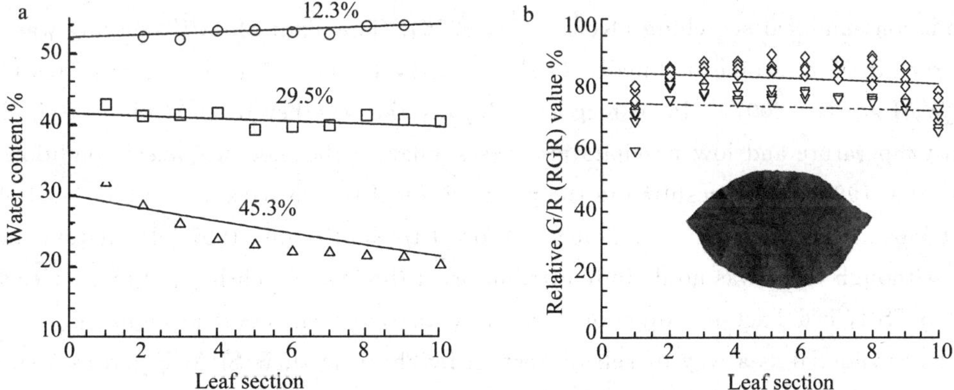

Fig. 8 Responsive curves of water content for isolated leaves with different water loss percentages, 12.3% (○—○), 29.5% (□—□) and 45.3% (△—△), to the leaf sections with different distances from the leaf proximal in Figure 8a. The leaf sections in the figure were numbered from proximal to distal of the leaf. It can be seen that detached leaves also dried from distal to proximal in 8a (line 45.3% water loss). Figure 8b is the responsive curves of RGR value for normal leaves (◇—◇) and natural dried detached leaves (▽—▽).

Under certain environment, a given plant usually grows by a pattern of logistic function accompanying with leaf area enlargement. The SPACF correspondingly enlarges gradually and is consistent with the around environment. According to the concept of cohesion tension (Kramer, 1983), water should cohere from root, stem and branch, and form a gradient to lift water to high elevation. Losing water balance results in a decrease in water potential that is transmitted to the shoots and leaves, and brings about closure to stomata to break the SPAC and reduce transpiration (Kramer, 1983). It is especially true for many evergreen tree species because they usually possess water-protective leaf structure (Schreiber *et al.*, 1996) or wax covered cuticle. Their cuticle transpiration is less than those deciduous and herbs. Under the prolonged drought environment the water loss through cortex still exists especially for the species with grass-like leaves. The isolated Kousa dogwood leaves lost water faster than those tree species with leathery leaves such as evergreen sasanqua, which lost water quite slow when stomata closed (Fig. 9). The SPAC cannot be easily broken thoroughly for these kinds of tree species that have high cuticle transpiration even though stomata closes. It is usually much easier for them to shear their SPACF under extreme droughty environment. For example, many deciduous tree species drop partial of their leaves by formation of separation layer and protective layer (Kozlowski, 1973; Fahn, 1990) at the base of petiole. In the persistent drought environment, even evergreens also appear the characteristics of SPACF shearing. Historically, in the early 1930's severe drought at the central states of United States and the unusual doughty weather in Australia in 1965, many trees appeared early defoliation, and leaf scorching (Kozlowski, 1976). The plant SPACF shearing was also observed in Mediterranean climate (Orshan, 1954, 1989) and in the limited sites (Van der Werf *et al.*, 2007). In Yamaguchi, Japan, the weather in 2007 characterized by high temperature and low precipitation was similar to the meteorological condition occurring in 1994, which resulted in the paddy rice leaf scorch-back and low production in west Japan (Yamamoto *et al.*, 1996) and forest trees (Kotani, 1997) defoliation and so on. Although there was no definite attribution of the leaf scorch-back of many tree species to TSR, it did act as a function partially withering their leaves as annual grasses. It can be considered as a way of self-protection by shearing their SPACF and as a special pattern for Kousa dogwoods to reduce their transpiring surface area.

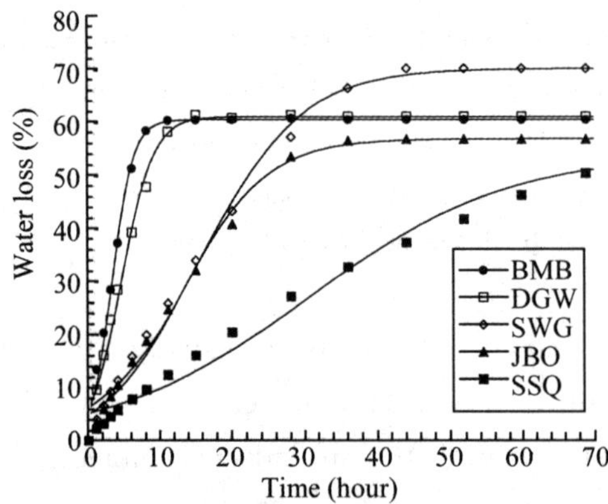

Fig. 9 The water loss process for detached leaves of five distinguished landscape tree and shrub species. In the figure, BMB is Kumazasa bamboo, DGW is Kousa dogwood, SWG is sweetgum, JBO is the Japanese blue oak and SSQ stands for sasanqua.

3.5 Triggering weather condition for TSR of Kousa dogwood

The SPACF shearing occurs when water absorption cannot keep up with transpiration losses seriously. It occurred on many Kousa dogwood trees during the abnormal extreme weather in 2007. The meteorological environment in 2007 in Yamaguchi, Japan was characterized by drought and hotness. The annual precipitation was lower and uneven, which was 71.6% of normal year and only 60.1% for the first nine months. The annual mean temperature accounted for 107.3% of the normal year. Particularly, it also made many new records of high temperature and low precipitation during past 41 years (from 1967 to 2007) for Yamaguchi meteorological observatory. The protracted drought led to insufficient soil water supply to plants, especially during sudden hot and droughty days. For example, in May and August in 2007, two peak periods of the AD11 in Yamaguchi from April to August (Fig. 10a) were observed and they were consistent well with the peroid of Kousa dogwood leaf scorch-back (Fig. 1d). During the first peak peroid of sudden increasing of the aridity, the accumulative precipitation from April to Jun. was only 47.4% of the normal and the precipitation from May 12nd, 2007 to Dec. 6th, 2007 was only 17.6% of total precipitation from April to Jun. In the second peak period, about 15 days' no rain and persistent high temperature and low humidity as well as high wind speed led to a foehn-like weather. Although the precipitation reached a height of making new record of maximum monthly precipitation in December, 2007 in Yamaguchi, it was late and out of the growing season. Various symptoms appeared on many landscape trees, especially the trees planted on the poor soils and the site with root growing limitation etc. During the special weather event of high temperature and less precipitation in 2007, large amounts of Japanese red pines (*Pinus thunbergii L.*) in mountain area died. Many tree species dropped partial leaves to reduce the transpiring surface. Some sasanqua trees shed all of their leaves to evade the extreme hot and droughty environment.

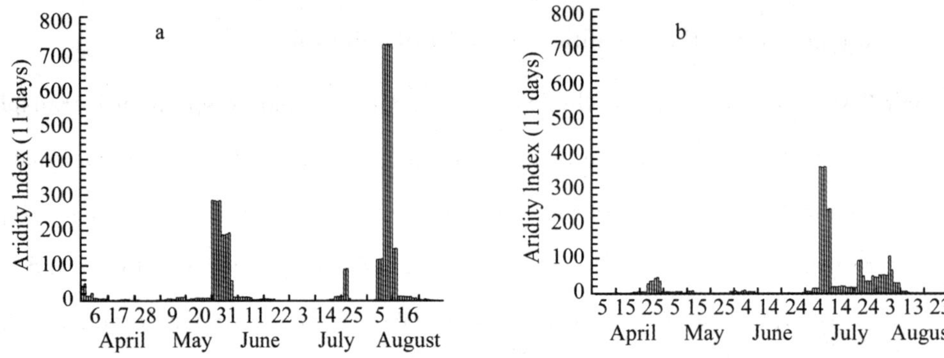

Fig. 10 10a gives the thirteen days' moving average aridity (AD11) from April 1st to the end of August in 2007. It clearly presents two peak periods of AD11. One occurred during the last ten days in May and the other during the middle ten days in August. 10b is the fifteen days' moving average aridity (AD11) index from April 1st to the end of August in 2008. It also clearly presents two peak periods. One occurred at the beginning of May and the other from July 4th to August 4th.

In 2008, sufficient precipitation during the first half year made the dogwood trees appear different responses from that in 2007. Some newly complementarily planted trees (it seemed that no perfect root system had been established), showed the gradual leaf scorch-back symptoms. The leaf scorch-back also occurred during two sudden hot and dry peak periods (Fig. 10b). For the previously planted Kousa dogwood trees no similar leaf scorch-back occurred until August 2nd after almost one-month's persistent drought and high temperature and it was also consistent well with a persistent peak peroid of AD11 in 2008 (Fig. 10b). Although the weather condition and water status of the Kousa dogwood trees varied, they all occurred during the sudden hot and dry period (Figs. 10a, 10b) and less water supplies in common. The sudden increasing of the aridity under the less water supply condition may be the triggering factor for the leaf scorch-back of the dogwood trees during the dual events of transpiration surface reduction in 2007 and 2008. So it may be usable to spray water on the surface of leaves or deep water the trees before serious drought and hot event occurrence to decrease the extreme water stress and maintain the normality of the transpiring cooler system as well as support more SPACF to dogwood trees.

References

[1] Adamsen F. J., Pinter P. J. Jr., Barnes E. M., LaMorte R. L., Wall G. W., Leavitt S. W., Kimball B. A. Measuring wheat senescence with a digital camera[J]. Crop Science, 1999, 39: 719-724.

[2] Addicott F. T. Physiological ecology of abscission, in "Shedding of Plant Parts"(Kozlowski T. T., ed.)[M]. New York: Academic Press, 1973: 103-104.

[3] Addicott F. T. Abscission[M]. London: University of California Press, 1982: 205-207.

[4]Bhat By. K. V., Sunrendran T., Swarupanandan K. Anatomy of branch abscission in Lagerstroemia Macrocarpa Wight[J]. New phytol, 1986, 133: 177-183.

[5] Clements H. F. Significance of transpiration[J]. Plant Physiology, 1934, 9: 165-172.

[6] Fahn A. Plant Anatomy (4th edition)[M]. Oxford: Pergamon Press, 1990: 262-263.

[7] Fitter A. H, Hay R. K. M. Environmental Physiology of Plants[M]. Academic Press, 2002: 162-169.

[8] Gunthardt-Goerg M. S., Vollenweider P. Linking stress with macroscopic and microscopic leaf response in trees, new diagnostic perspectives[J]. Environmental Pollution, 2007, 147: 467-488.

[9] Kotani E. The stand structure and diameter increment of Hinoki cypress after hit by summer drought caused by less rainfall in 1994[J]. Application Research of Forest, 1997: 25-28. (In Japanese)

[10] Kozlowski T. T. Shrinking and swelling of plant tissues, in "Water Deficits and Plant Growth" (Kozlowski, T. T., ed. 3)[M]. New York: Academic Press, 1972: 1-64.

[11] Kozlowski T. T. Shedding of Plant Parts[M]. New York: Academic Press, 1973: 1-117.

[12] Kozlowski T. T. Water supply and leaf shedding, in "Water Deficits and Plant Growth" (Kozlowski T. T., ed. 4)[M]. New York: Academic Press, 1976: 191-222.

[14] Kramer P. J. Water Relation of Plants[M]. Academic Press, 1983: 187-213.

[15] Miura T., Ono T. Classification of Toprography in "Yamaguchi-Ogori Part of Fundamental Survey of Land Classification in Japan"[M]. 1972: 14-15.

[16] Orshan G. Surface reduction and its significance as a hydroecological factor[J]. J. Ecol, 1954, 42: 442-444.

[17] Orshan G. Plant phono-morphological studies in Mediterranean type ecosystems[M]. Kluwer Academic Publisher, 1989: 398-399.

[18] Rust S., Roloff A. Acclimation of crown structure to drought in Quercus robur L. Fintra- and inter-annual variation of abscission and traits of shed twigs[J]. Basic and Applied Ecology, 2004, 5: 283-291.

[19] Sakaue Y., Ijiri T. Soil interpretation in "Yamaguchi-Ogori Part of Fundamental Survey of Land Classification in Japan"[M]. 1972: 32-33.

[20] Schreiber L., Riederer M. Ecophysiology of cuticular transpiration: Comparative investigation of cuticular water permeability of plant species from different habitats[J]. Oecologia, 1996, 107: 426-432.

[21] Thornley J. H. M. Mathematical Models in Plant Physiology[M]. Academic Press, 1976: 48-50.

[22] Treshow M. Environment and Plant Response[M]. Mcgraw-Hill Publications in the Agricultural Science, 1970: 22-34.

[23] Van der Werf G. W., Sass-Klaassen U. G. W., Mohren G. M. J. The impact of the 2003 summer drought on the intra-annual growth pattern of beech (Fagus sylvatica L.) and oak (Quercus robur L.) on a dry site in the Netherlands[J]. Dendrochronologia, 2007, 25: 103-112.

[24] Vollenweider P., Guünthardt-Goerg M. S. Erratum to "Diagnosis of abiotic and biotic stress factors using the visible symptoms in foliage"[J]. Environmental Pollution, 2006, 140: 562-571.

[25] Yamamoto H., Suzuki Y., Hayakawa S., Hirayama K. Survey on meteorological characteristics of dry summer and paddy rice damage caused by the drought in western part of Japan in 1994[J]. Journal of Natural Disaster Science, 1996, 15: 11-17.

[26] Yapp R. H. Spiraea Ulmaria and its bearing on the problem of xeromorphy in marsh plants[J]. Ann. Bot., 1912, 26: 815-870.

(原文发表于 *Journal of Forestry Research*, 2009, 20(4): 337-342)

Image measurements of leaf scorches on landscape trees subjected to extreme meteorological events*

WANG Fei[1], Kenji Omasa[2]

(1. Shandong Forestry Research Academy, Jinan, China; 2. Graduate School of Agriculture and Life Sciences, the University of Tokyo, Japan)

1 Introduction

Extreme meteorological events, such as strong typhoons and summer droughts, in addition to root disease often induce tip and/or margin scorches on trees (Yapp, 1912; Kozlowski, 1976; Gunthardt-Goerg, Vollenweider, 2007; Vollenweider, Gunthardt-Goerg, 2006; Wang *et al.*, 2009a, 2009b) due to severe dehydration. A variety of die-back and decline diseases of trees are favored by abiotic stresses, such as drought, flooding, extreme heat, mineral deficiency and air pollution. These diseases usually become severe after environmental stresses render trees susceptible to organisms of secondary action (Kozlowski, 1985; Omasa, 2002; Schoeneweiss, 1981). Moreover, myxomycetes often confine their attacks to the same regions of a leaf (Yapp, 1912). Leaf scorches often show systematic symptoms when the trees are subjected to severe water stresses. Similar symptoms may also be induced by several biotic agents, viruses and insects (Hammerschlag *et al.*, 1986). However, pathogen-triggered injuries generally cause unsystematic necrosis in a localized pattern and on partial crowns of trees and shrubs. Leaf scorch symptoms originating from extreme environmental events often appear in an inward spreading pattern. A type of inward leaf scorch symptom on dogwood trees has been found in Yamaguchi, Japan, which was hit by a summer drought in 2007. (SD2007) (Wang *et al.*, 2009b) To identify the difference between abiotic leaf scorch and biotic damage, the parameters of internal angle of injured area (IAIA) and percent

* Supported by the National Natural Science Funds of China (ID Number: 301170671).

of scorched central vein (PSCV) were defined. The IAIA is a central angle corresponding to the scorched leaf margin, and the PSCV is a length rate between the scorched part and the total length of the central vein in accord with the percent of scorched leaf tip. They respond the scorch symptoms with respect to their spreading from both radian and length.

Image analysis have increasingly become important tools to holistically study the characteristics of plants or trees (Omasa, 1990; 2002; Adamsen *et al.*, 1999, 2000; Omasa *et al.*, 2007, Jones 2004; Wang *et al.*, 2008a, 2008b; Richardson *et al.*, 2001). RGB image analysis has been recognized as a useful tool to determine the leaf scorch area percent (Wang *et al.*, 2009b; Zhao *et al.*, 2012), and it has also been used to directly measure the heterogeneousness of leaf living parts and dead parts (Wang *et al.*, 2009b). The common parameters of image analysis, such as distance, angle and area measurement, are useful parameters to describe the morphology of plants and trees (Bai *et al.*, 2005; Bréda, 2003). In this study, it is aimed at objectively responding the ecological information from some landscape trees and finding an easy and low cost way to diagnose the leaf scorches suffered extreme meteorological events by using the digital image analysis. Another goal is focused on the proper parameter or index to describe the scorch symptoms. As a result, the IAIA value for leaf scorching, spot/anthracnose diseases and wind mechanical damage as well as the PSCV for scorched dogwood leaves etc. were determined by using the software of UTHSCSA Image Tool 3.00 to investigate the difference among the symptoms caused by abiotic stress, biotic attack and mechanical damage. The positive result suggests that leaf scorch of these tree species is a systematic symptom which may be potentially diagnosed by using RGB image measurement.

2 Materials and Methods

2.1 The studied position and tree species

This study was mainly carried out at Yamaguchi, Japan. The target landscape trees and shrubs included the following species: kousa dogwood (*Cornus kousa* Buerg.), which suffered severe SD2007 at Yamaguchi Central Park; Japanese blue oak (*Quercus glauca* Thunb.) and camphor tree [*Cinnamomum camphora* (L.) J. Presl.], which were struck by the strong dry typhoon 0613 (T0613); trident maple (*Acer buergerianum* Miq.) and zelkova (*Zelkova serrata* Murr.), which were hit by strong spring wind at Yamaguchi University; and Japanese spindle (*Euonymus japonicus* Thunb.), winter creeper ([*Euonymus fortunei* (Turcz.) Hand.-Maz.], Glossy Privet (*Ligustrum lucidum* Ait.) and red leaf photinia (*Photinia glabra* (Thunb.) Maxim.) supplement shrubs in several hedges, which suffered winter injury at the Shandong Forestry Research Academy in Jinan, China, and so on.

2.2 Preconditions and methods of image taking

Leaves were monitored before and after the symptom appearance of leaf partial withering, disease and wind mechanical damage to confirm the origin of the symptoms. Leaf scorch events were recorded soon after symptoms appearance, when the leaf lesion color still had the same color as the living part. During study of the partial freezing dehydration process of some broad evergreen tree species such as Japanese spindle, glossy privet and so on, the freezing and thawing process of their leaves were monitored.

The leaves were sampled from sampling trees and took photos with a CCD camera (Canon IXY 6.0) or scanned with a flat bed scanner (Canon d125u2). Images were stored in the form of Tiff format. There is no special constraint for the photograph taking in manual measurement of IAIA, PSCV and LSAP except the blurry images of leaf partial withering symptom for most tree species. Images will be increased luminance by using the adjust tool in Photoshop software if the object leaf is too dim to be seen clearly. The number of sampled leaves for each tree species is shown in the bracket following their names in the figure captions (From Fig. 3 to Fig. 6).

In order to increase the sample numbers of diseased and scorched leaves and to check multiple tree species, a large number of photographs from books and publishing papers have also been measured.

2.3 The definition and calculation of IAIA and PSCV values

The internal angle of injured area (IAIA) is a central angle when considering the farthest point that the injury has reached on the central vein (or in the injured area if the injured area is beyond the central vein) as tip point of the angle and considering the two lines from this point to the farthest points that the injury reached on both sides of leaf edge as the two borders of this angle. (Figs. 1b and 1c) The IAIA was directly measured by using the angle analysis tool in the software of UTHSCSA Image Tool 3.00.

The PSCV was measured by using the distance tool in the software of UTHSCSA Image Tool 3.00 and calculated using Equation (1) as follows:

$$PSCV = (CV/TV) \times 100\% \quad (1)$$

in which, PSCV is the percent of scorched central vein, CV (from C to V in Fig. 1a) is the pixel number of scorched part and TV (from T to V in Fig. 1a) is the pixel number of the total length of the central vein.

2.4 Process and method to determine the LSAP

The target kousa dogwood trees were approximately 7 years old and 3 to 4 meters high, and these trees were planted around the Yamaguchi Central Park. The number of sampled trees totaled 60. In this study, the leaf scorch area percent (LSAP) values of the sampled leaves were determined by image analysis. All of these trees were monitored and analyzed by image measurement during the years of 2007, 2008 and 2009. In this study, 560 leaves were scanned with a flat bed scanner (Canon d125u2). Photoshop CS 12.0 software was used to remove all objects except the leaves from the images. The scorched area and overall leaf area was measured by using area tool in the UTHSCSA Image Tool 3.00 software or directly read the pixels from Photoshop. The LSAP was then calculated using Equation (2) as follows:

$$\mathrm{LSAP} = 100 - (100 \times \mathrm{PGAL}/\mathrm{POL}) \tag{2}$$

where PGAL is the number of pixels in the green area of the leaf, and POL is the number of pixels in the overall leaf. (Fig. 1a)

A multivariate normality test has been carried out for the data of IAIA, LSAP and PSCV, they are all consistent with the normal distribution with the U checking value less than 1.96 and $p>0.05$.

2.5 Meteorological data and related parameter

The precipitation data were obtained from the Automated Meteorological Data Acquisition System (AMeDAS) in Japan. The three month accumulated precipitation (TMP) was calculated with Equation (3) to compare the seasonal difference in precipitation among the years 2007, 2008 and 2009 as follows:

$$\mathrm{TMP}_i = \sum_{j=0}^{2} \mathrm{MPR}_{i+j} \tag{3}$$

where $i=1, 2...,10$ and $i=1$ in January; $j=0$, 1 and 2; MPR_{i+j} is the monthly precipitation value in the month of $i+j$.

2.6 Leaf venation, water-trail and red-color-trail observations

Leaf venation of target tree species has been studied by using both methods of micro-distance photography and chemical treatment. In the study, most tree species with transparent leaflets were directly observed by using micro-distance photograph, while the tree species with thick and nontransparent leaflets were first treated with NaOH solution to obtain the clear venation images.

By monitoring the leaflet of some evergreen arbor/shrub species, the water-trail, a phenomenon of partial leaflet temporary water deficit with significant visible trace during the process of occasional winter frozen, was studied. Then, the red-color-trail has been observed at the same area where water-trail has been found. They are all similar to the permanent damage symptoms of leaf scorches after extreme winter dehydration.

Therefore, the IAIA of them were measured with similar method to prove that they also result from the water metabolism and imbalance of water system.

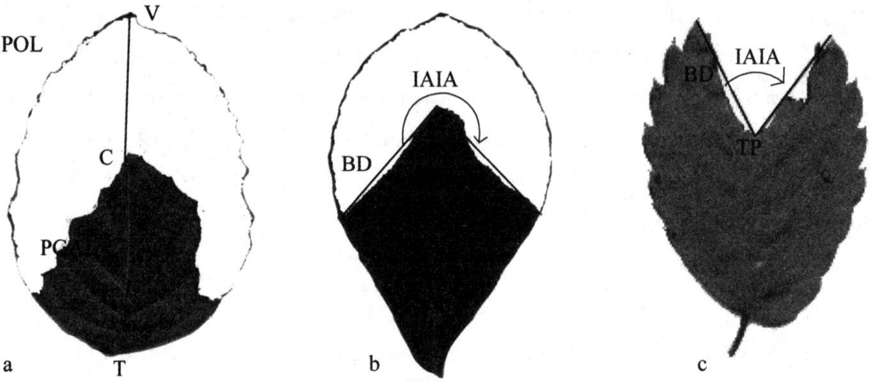

Fig. 1 Diagrams to show the internal angle of injured area (IAIA) and leaf scorch area percent (LSAP). (a) A typical scorched kousa dogwood leaf with the non-scorched green area (PGAL) and entire leaf area (POL) marked. (b) A scorched Japanese spindle leaf with a typically obtuse IAIA. (c) A zelcova leaf mechanically damaged by wind with a significant acute IAIA. The IAIA was measured by the tip point (TP) and side borders (BD).

3 Results and Analysis

3.1 Extreme meteorological events in 2007, 2008 and 2009 and responses of several kousa dogwood trees in Yamaguchi

In the presence of meteorological changes in the world, the climate in Yamaguchi, Japan fluctuated among the years 2007, 2008 and 2009. The fluctuations were mainly reflected in the three month precipitation values during these years. (Fig. 2a) As for the meteorological environment in 2007 in Yamaguchi, Japan was characterized by drought in almost all of the first eleven months and by high temperatures in February, August, September and October. The annual precipitation was 71.6% of the normal precipitation of this area, and the precipitation for the first nine months was only 60.1% of the normal precipitation. From August 7th to August 18th in 2007, more than 15 days of continual anticyclone weather occurred. During these days, the average and minimum temperatures were higher than normal (1971-2000), and no rainfall occurred.

Meanwhile, the relative humidity values represented a U-shaped curve. Although the precipitation during the first half of the year in 2008 was similar to normal years, the annual deviation of mean temperature in the summer was 0.5 ℃ higher than normal, and the precipitation was less than 80% of the normal precipitation at Yamaguchi. During the hottest days from July 22nd to August 14th in 2008, only 11 mm of precipitation was recorded at the Yamaguchi Observatory. The maximum daily temperature during this period was higher than 33 ℃. The characteristic precipitation in 2009 presented a different tendency compared to the years 2007 and 2008, especially in the summer. The prolonged Japanese rainy season and historical record created heavy rainfall in July caused local flooding in Yamaguchi. More than 750 mm of monthly precipitation led to a peak in the TMP as shown in Fig. 2a. After this period, less rainfall in September resulted in a record of the third least monthly rainfall in Yamaguchi meteorological history.

This strongly contrasting climate induced different responses from landscape trees in Yamaguchi, especially for the kousa dogwood trees planted in the Yamaguchi Central Park. (Fig. 2b) According to the LSAP values of the 60 fixed sampling trees, more than 40% of the leaf area was reduced by partial withering in 2007. The leaf scorch area of 13% in 2008 for the same sampling trees may be a result of the normal precipitation in the first half of the year. (Fig. 2a) In 2009, the LSAP was only 0.3%, which implied that almost no leaf scorch occurred. According to the variation analysis, LSAP values varied significantly among 2007, 2008 and 2009 with an F-value equal to 62.6 ($p<0.01$). There was a consistency between the leaf scorch area percent and precipitation during the summer days, especially in July, August and September. The LSAP indirectly indicated that the heavy rain in the summer of 2009 counteracted the summer heat wave and caused almost no leaf scorch occurrence on the kousa dogwood trees planted in Yamaguchi Central Park. Similarly, the largest IAIA value for kousa dogwood leaves was obtained in 2007, followed by the IAIA values in 2008 and 2009. A significant difference in the IAIA values was measured among the years with an F-value of 144.76 ($p<0.01$). (Fig. 2c)

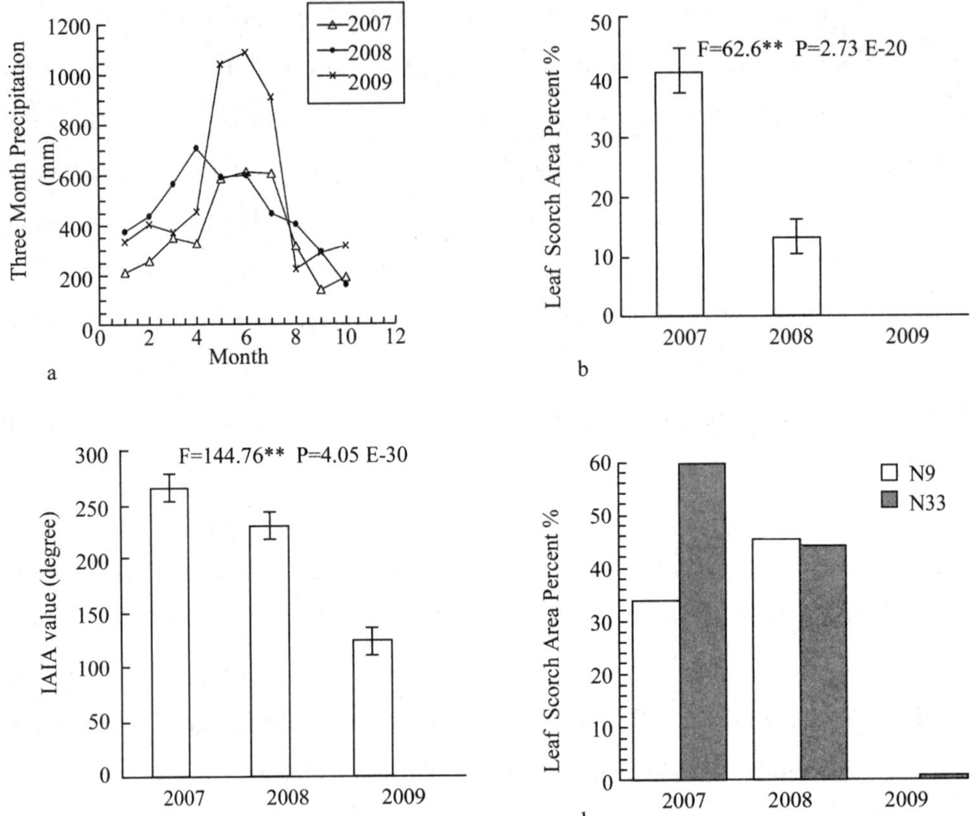

Fig. 2 Precipitation characteristics and responses of several kousa dogwood trees in Yamaguchi in 2007, 2008 and 2009. (a) Three month precipitation (TMP) during 2007, 2008 and 2009. (b) Leaf scorch area percent (LSAP). (c) internal angle of injured area (IAIA) for 60 kousa dogwood sample trees planted around the Yamaguchi Central Park. (d) Leaf scorch area percent values in 2007, 2008 and 2009 for two kousa dogwood sample trees (number 9 and number 33) with leaf scorch present in all of the three years.

There was almost no appearance of leaf scorch symptoms by August 28th in 2009, which was the date of investigation in all three years of the study period. Leaf scorch symptoms among the kousa dogwood sample trees did not occur until the end of September in 2009, which may have been attributed to the lower incidence of rainfall in September of 2009. According to the investigation, only two individual trees had leaf scorch symptoms on September 20th after prolonged lack of rain and high temperatures. By comparison, the LSAP of these two individual trees (N9 and N33) in 2009 was far lower than that in 2007 and 2008 and could be considered nonoccurrence as compared to 2007 and 2008. (Fig. 2d) Moreover, this result may provide further evidence that the leaf scorch of kousa dogwood trees in Yamaguchi Central Park is associated with the water stress induced by less rain and high temperatures during summer.

3.2　The LSAP, IAIA and PSCV of scorched kousa dogwood leaves

Careful observation of kousa dogwood leaves reveals that leaf veins are thinner from the base to the tip and from the central vein to the margin. Similarly, leaf scorch caused by extreme meteorological events usually starts from the leaf tip and margin of kousa dogwood leaves. This phenomenon to some extent also appeared in this study by measuring the IAIA and PSCV, which reflect the radian and length of the scorched area, respectively, of kousa dogwood leaf images. Thus, as the IAIA and PSCV values increased, the tip and margin scorches became more severe.

According to the measurement of these two parameters, a tendency of the IAIA to increase as the LSAP increased (or a significantly positive exponent relationship) was found (Fig. 3a), with an R^2 value of 0.556 ($p<0.01$). Moreover, a statistically significant and positive exponent relationship existed between the PSCV and LSAP values (Fig. 3b), with an R^2 value of 0.841 ($p<0.01$). Thus, as the percent of scorched leaf area increased, a larger area of the leaf tip and margin died of meteorological injury. Therefore, the kousa dogwood leaves showed leaf scorch symptoms starting from the tip and margin of the leaf.

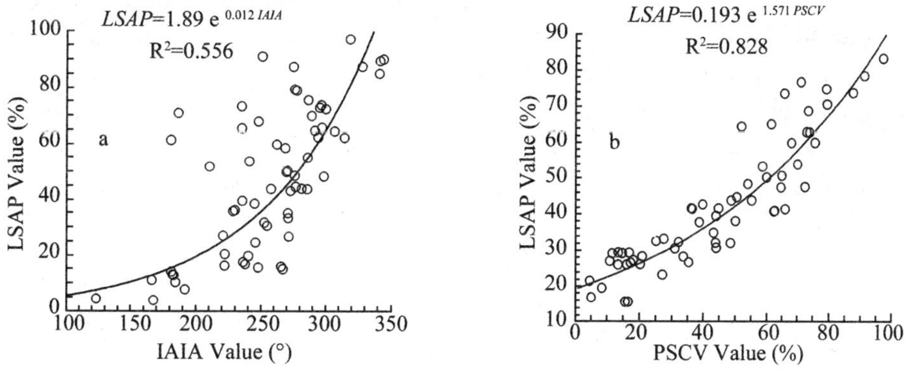

Fig. 3　Relative curves of (a) leaf scorch area percent (LSAP) versus internal angle of injured area (IAIA) and of (b) LSAP versus percent of scorched central vein (PSCV) ($n=65$).

3.3　The IAIA value of frozen dehydrated leaves with water-trail and red-color-trail

Persistent winter frozen dehydration induced similar symptoms to leaf scorches, such as water-trail (Figs. 4a-1, 4a-2, 4a-3 and 4a-4) and red-color-trail (Figs. 4a-5, 4a-6) on leaves of some landscape trees with larger IAIA values (Figs. 4b-1, 4b-2, 4b-3, 4b-4, 4b-5 and 4b-6) of more than 250°. The water-trail is usually restorable and disappears after thaw of the frozen leaves. Permanent leaf scorch often occurred at the same area as the frozen dehydration persisted for a longer time (Fig. 4a-2, tip), during severe drought or at the extreme stressed environment. For example, in the winter from 2011 to 2012 only a few Japanese spindles around the yard of Shandong Forestry Research Academy in Jinan, China appeared permanent symptoms of leaf scorch starting from tip and margin, in which some of them were new planting seedlings with imperfect root system and some of them grew on new branches sprouting from an old root stock. Whatever, there has been a significant positive relation between water-trail area rate (W-rate) during frozen dehydration and the green area rate (G-rate) of these leaves after permanent scorch (Fig. 4c). It suggests that the water and energy imbalance become the trigger of permanent scorch, the water-trail and red-color-trail symptoms, which start from the leaf tip and margin.

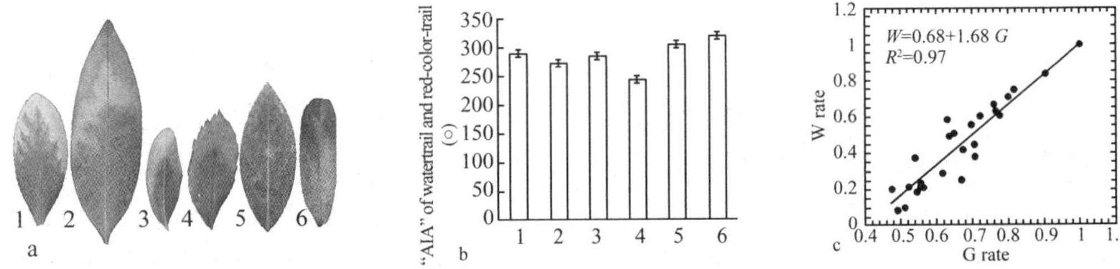

Fig. 4 The images of leaf water-trail (a, 1-4) and red-color-trail (a, 5-6), their IAIA value (b) of some landscape tree species and the relation between water-trail area rate (W-rate) of frozen dehydrated leaves and green area rate (G-rate) of permanently partial withered leaves of Japanese spindles (c) during or after the process of frozen dehydration, in which 1 is the Japanese spindle (n=63), 2 the glossy privet (n=26), 3 the purpus privet (n=36), 4 the wichura rose (Rosa *wichuraiana* Crep.) (n=11), 5 the winter creeper (n=13) and 6 the red leaf photinia (n=18).

3.4 Comparison of multiple leaf partial injuries of several landscape trees

Direct image measurement of leaf wind break was seldom found. In 2009, strong windy weather accompanied by heavy rainfall occurred twice in Yamaguchi, in late spring and early summer (on April 20th and May 21st). The maximum gust wind speeds reached 26.2 m/s and 19.9 m/s on April 20th and May 21st, respectively. Due to the strong wind on April 20th, severe mechanical damage was found on the juvenile leaves of zelcova, Japanese blue oak and trident maple trees, and the severity of the damage was also attributed to the early leaf stage and leaf textures. One month later, however, the same intensity of strong wind had no influence on the leaves of these tree species because the leaves were more developed and stronger, which allowed them to endure the strong wind. The wind break damage to the leaves occurred mainly on juvenile leaves and seldom on mature leaves. The injured juvenile leaves usually had smaller IA-

IA values (Figs. 5a-3, 5a-4, 5a-5, 5b-3, 5b-4 and 5b-5) similar to the anthracnose-diseased dogwood leaves (Figs. 5a-1, 5a-2, 5b-1 and 5b-2). However, the strong and hot wind during the summer drought period in August of 2007 triggered leaf scorch symptoms in most of the kousa dogwood trees planted around Yamaguchi Central Park. The IAIA values of kousa dogwood leaves after the summer drought period in 2007 were greater than 250° (Figs. 5a-7 and 5b-7). In 2006, many landscape trees had leaf scorch on the windward side of their crowns after being hit by strong wind caused by T0613 accompanied with less rainfall. Some camphor trees and Japanese blue oaks were exceptional examples with large IAIA values (Figs. 5a-6, 5a-8, 5b-6 and 5b-8). A similar mechanism was not found simultaneously for both mechanical damage on younger leaves and leaf scorch caused by extreme meteorological events. Interestingly, during the winter injury process, with the average monthly temperature from November 2009 to March 2010 being 1.4 ℃ lower than normal, some evergreen shrub species planted in Jinan City, China presented similar symptoms to those hit by extreme meteorological events during SD2007 and T0613 in Yamaguchi, Japan with larger IAIA values (Figs. 5a-9, 5a-10, 5a-11, 5b-9, 5b-10 and 5b-11). Therefore, the symptoms may be related to water stress as evidenced by the similar symptoms after suffering from the extreme meteorological events of SD2007 and T0613. Winter injury has ever been considered to be the result of a frozen dehydration process (Pearce, 2001), especially in areas where there is a strong winter wind (Pirone, 1978). The IAIA values of these evergreen tree species were similar to the IAIA values of leaf scorch on the kousa dogwood trees caused by summer drought. Thus, these results suggested that the frozen dehydration process has similar mechanism to desiccation or strong, dry typhoon strikes.

Fig. 5　Typical symptom (a) and internal angle of injured area (IAIA, b) of some landscape trees. (1) IAIA of leaf injury to kousa dogwood leaves caused by anthracnose disease (n=5) in Yamaguchi, Japan. (2) IAIA of leaf injury to kousa dogwood leaves caused by both water stress and secondary disease (n = 26) in Yamaguchi, Japan. IAIA of leaf injury to (3) Japanese blue oak (n=35), (4) zelcova (n=28) and (5) trident maple (n=4) leaves caused by strong wind during the leaf expanding stage in April 2009 in Yamaguchi, Japan. IAIA of leaf injury to

(7) kousa dogwood leaves (n=25) triggered by SD2007, (6) IAIA of leaf injury to Japanese blue oak leaves (n=31) and (8) camphor tree leaves (n=34) induced by T0613 in Yamaguchi, Japan. IAIA of leaf injury to (9) Japanese spindle (n=26), (10) winter creeper (n=23) and (11) red leaf photinia (n=6) leaves caused by severe winter frozen dehydration in Jinan City, China.

Theoretically speaking, the limitation value of the internal angle of the leaf spot disease should be the limitation of the circumferential angle of a circle (or 180°), which is equal to half of the limitation of the central angles of a circle (360°). Therefore, it can be deduced that the IAIA value of an angular leaf spot disease of dogwood is less than 180°.

3.5 Quantitatively measuring results of leaf scorch from historical reports

The IAIA of leaf images measured by the past reports is an easiest way to study the leaf scorch symptoms. By measuring the typically scorched leaves from the reports of Yapp (1912), the larger IAIA of all leaves tally with this conclusion even for the linear leaf of Buckhorn Plantain (*Plantago lanceolata L.*) (Fig. 6A; IAIA=205.9). The IAIA of the typically sampling leaves for the tree species of Common Oak (*Quercus robur L.*), Europea Beech (*Fagus sylvatica L.*), European-haw (*Crataegus oxyacantha*), poplar (*Populous Spp.*) and rose (*Rosa Spp.*) exposed to strong wind were even more than 300 degree (Fig 6. B, C, D, E and F). Similarly, all leaves from America linden (*Tilia spp.*), buckeye (*Aesculus carnea*), kousa dogwood (*Cornus kousa Buerg.*) and American elm (*Ulmus Americana L.*) suffered environmental stresses, with partial withering symptoms from the report of Schoeneweiss (1981) which also showed larger IAIA more than 300 degree (Fig 6. G, H) except the American elm leaf suffered bacteria leaf scorch (Fig 6. K; IAIA=76.9).

According to the measurement of three groups of leaf spot disease and anthracnose disease from some plant disease books consisting of 16, 53 and 57 cases (Ito, Aino, 1976; Pirone, 1978), the average IAIA values respectively equal to 119.7±22.8°(Fig 6. YT), 142.0±9.8°(Fig 6. ZG) and 106.9±8.3°(Fig 6. BD). These results were in accordance with the disease or mechanical damage samplings shown in Fig. 6. They were far less than the mean IAIA values of leaf scorch symptoms caused by SD2007, T0613 and prolonged winter injury with average IAIA values ranging from 250° to 320° with significant statistical differences due to the F value of 63.5 and p-value 6.58e−67.

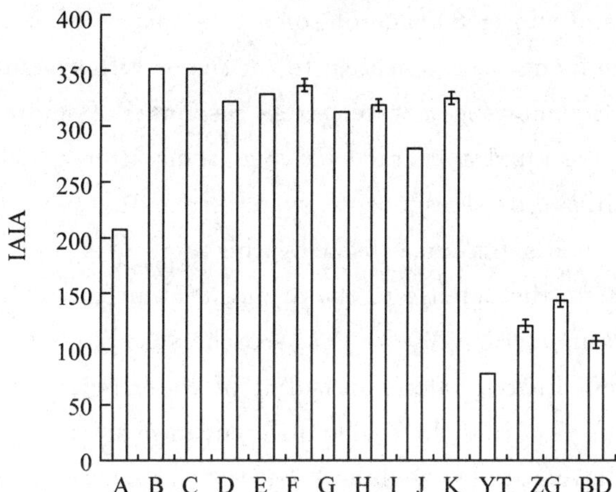

Fig. 6　IAIA value measured from the reports on leaf scorch and diseases. It includes leaf scorch on the linear leaf of Buckhorn Plantain (A), on the leaves of Common Oak (B), Europea Beech (C), European-haw (D), Poplar (E), Rose [F (n=5)], America Linden (G), Buckeye (H), Kousa dogwood (I) and American elm [J (n=17)]; bacteria leaf scorch (Bls) on American elm (K); and the anthracnose/spot disease symptoms from different plant disease books [Yt (n=16), Zg (n=53) and Bd (n=57)].

4　Discussion

Exogenous environmental factors become operative through the endogenous metabolic processes of trees. The origination and spread of disease symptoms also can be considered as an interaction process between pathogen intrusion and plant resistance. Pathogen spreading is a complex action, which includes the intrusion of a hypha into host in the pattern of cell by cell. A hypersensitive response from the host usually confines the pathogen into a localized necrotic lesion even if the pathogen fails to further intrude into the host if it is incompatible. This type of host defense action varies among cultivars with different sensitivities to the pathogen (Wharton *et al.*, 2001) or varies between attached leaves and detached leaves (Liu *et al.*, 2007). This kind of leaf-pathogen interaction even can be presympotmatically detected by using thermographic technique (Chaerle *et al.*, 1999; Omasa *et al.*, 2002; Jones, 2004; Wang, 2010). Therefore, the necrotic lesion on a diseased leaf is usually impeded by the living cells all around if the intrusion point is located just within the leaflet. The spot/anthracnose disease induced by fungus as well as the mechanical damage induced by strong wind occurring on the juvenile leaves of some tree species had relatively smaller IAIA values because the symptom spread was often resisted from all directions, except for the leaf margin in the case of margin intrusion.

By comparison, the leaf scorch caused by severe meteorological extreme events such as summer drought and winter damage of some tree species often presented the resistance coming from the living tissues nearest to the major vein system. Kozlowski (1976) considered that visible symptoms of water stress responses of landscape trees often show temporal delay, and the rapid appearance of symptoms after rehydration suggests that the symptoms are initiated by a response to water stress that cannot be completed without adequate water. The symptom caused by this type of systematic shock often has characteristics similar to the defense to the drying back progress of the tissues furthest from water-way. (Wang *et al.*, 2009a) Leaf scorch caused by water stress appears as a pattern that constructs a barrier along an isoline of water potential to protect from further water loss (Wang *et al*, 2009b). The different mechanisms of these stresses cause distinct symptoms. In this study, we found that the leaf scorch originating from desiccation usually had large IAIA and PSCV values due to the thinner veins at the tips and margins of leaves. Thus, this result supports the hypothesis that leaf scorch often starts from the leaf parts furthest from water-way (Yapp, 1912; Gunthardt-Goerg, Vollenweider, 2007), and it also suggests that image analysis may be a useful tool for comparing leaf scorch symptoms caused by desiccation and spot/anthracnose disease.

Leaf scorches often occurred on transplanting seedling/trees, shade trees (Hammerschlag *et al.*, 1986; Kozlowski, 1985) and extreme environment planted trees, and on leaves of watershoot, which easily present water imbalance during striking by meteorological extremes. Some shade trees of Japanese blue oak subjected to a strong wind at Yamaguchi University presented leaf scorch on the windward side after T0613, which indicated that serious water imbalance occurred in the thin shade leaves facing the strong and dry wind. However, the remaining green parts of the scorched leaves maintained normal function for more than three years (Wang *et al.*, 2009a). By comparison, the IAIA values of these leaves were significantly different from the IAIA values of leaves that experienced mechanical damage caused by strong early spring wind in 2009 and similar to the IAIA values of leaves that were hit by summer drought events in 2007.

Some diseases and mechanical damage caused the leaves of landscape trees have similar symptoms with smaller IAIA values, which implied that the symptoms did not cause systemic damage in the trees. The leaf scorch symptoms caused by summer drought, dry typhoons and persistent winter frozen dehydration presented similar symptoms with larger IAIA values, which suggested that these results all triggered systemic variation through water metabolism. The recoverable symptoms of water-trail on some evergreen tree species during the severe frozen dehydration may be the actual evidence of its water imbalance. The phenomenon of red-color-trail appeared at the same area of water-trail suggests that it is the internal responses of trees to water stresses induced by extreme meteorological events.

In addition, the symptoms and injury mechanisms of polluted plants are usually different from the leaf scorch symptoms caused by extreme meteorological events (Mudd, Kozlowski, 1975; Omasa *et al.*, 1981). Therefore, there is potential to diagnose the symptoms of leaf scorch caused by extreme meteorological events from other damages (Vollenweider, Gunthardt-Goerg, 2006). In North America, bacterial leaf scorch is affected by both *Xylella fastidiosa* and water stress (McElrone *et al.*, 2001; Wood *et al.*, 1995). The dualism of their symptoms (McElrone *et al.*, 2001; Hammerschlag *et al.*, 1986) indicates that the diagnosing method described in this paper may be used for morphologically determining which one is the major factor between water stress and bacterial infection.

Although IAIA differences among distinct injuries are consistent, it remains unclear for the universality of most plant or tree species and it needs to be further proven by measuring a large number of samples. The judgment of injuries was complicated because abiotic stresses usually predispose the occurrence of disease (Kozlowski, 1985; Schoeneweiss, 1981; Desprez-Loustau *et al.*, 2006; Blodgett *et al.*, 1997), sometimes abiotic stress and disease coexisted on the same tree, and the responses varied with tree species. Although the diseased leaves usually showed local lesions at leaf margins with small IAIA, some lesions occasionally connected together, which made the lesions look like the typical characteristics of water stress, especially during the second half of the disease process. Anyhow, the typical disease characteristics can be statistically manifested when sample sizes are large enough in the early stage of the disease. In contrast, partial leaves from few water loss-resistant tree species, especially for some evergreen tree species with larger leaflets, sometimes presented local necrotic lesions away from the margin similar to disease symptoms. Tree species with these types of characteristics usually contained venations that lacked the veins with thin tip and margin or gradually thinner veins from base to tip and from central vein to margin. Therefore, measurements should be perfected before the universality of the symptoms on most plant or tree species is confirmed.

5 Conclusion

The IAIA, LSAP and PSCV become effective indices to respond the scorch symptoms of dogwood trees etc. induced by extreme meteorological events. Meanwhile, the larger IAIA values for scorched leaves and smaller IAIA values for spot/anthracnose diseased leaves of the studied trees indicate that IAIA may be used as a potential diagnosing index of leaf scorch symptoms after further studies. This study suggests that digital image analysis can be a useful and low-cost tool in distinguishing leaf scorch symptoms from disease and mechanical damage on some landscape trees. It makes the study of leaf scorch being possible from experiential comparison to quantitative and statistical analysis in the scope of eco-infomatics.

References

[1] Adamsen F. J., Pinter P. J. Barnes E. M. Jr., Lamorte R. L., Wall G. W., Leavitt S. W., Kimball B. A. Measuring wheat senescence with a digital camera[J]. Crop Sci., 1999, 39: 719-724.

[2] Adamsen F. J., Coffelt T. A., Nelson M., Barnes E. M., Rice R. C. Method for using images from a color digital camera to estimate flower number[J]. Crop Sci, 2000, 40: 704-709.

[3] Bai J. H., Wang K., Chu Z. D., Chen B., Li Sh. K. Comparative study on the measure methods of the leaf area[J]. *J. Shihezi Univ.* (*Nat. Sci.*), 2005, 23: 216-218.

[4] Blodgett J. T., Kruger E. L., Stanosz G. R. Sphaeropsis sapinea and water stress in a red pine plantation in Central Wisconsin[J]. Phytopathology, 1997, 87(4), 429-434.

[5] Bréda N. J. J. Ground-based measurements of leaf area index: a review of methods, instruments and current controversies[J]. *J. Exp. Bot.*, 2003, 54(392): 2403-2417.

[6] Chaerle L., Caeneghem W. V., Messens E., Lambers H., Montagu M. V., Van Der., Straeten D. Presymptomatic visualization of plant virus interactions by thermography[J]. Nat. Biotechnol., 1999, 17: 813-816.

[7] Chinese Academy of Agricultural Science (CAAS) (Eds.). Illustrations of Crop Disease and Insect Pests of China[M]. Beijing: Beijing Agricultural Press, 1965.

[8] Desprez-Loustau M., Marçais B., Nageleisen L., Piou D., Vannini A. Interactive effects of drought and pathogens in forest trees[J]. Ann. For. Sci., 2006, 63, 597-612.

[9] Gunthardt-Goerg M. S., Vollenweider P. Linking stress with macroscopic and microscopic leaf response in trees, new diagnostic perspectives[J]. Environ. Pollut., 2007, 147: 467-488.

[10] Hammerschlag R., Sherald J., Kostka S. Shade tree leaf scorch[J]. J. Arboric., 1986, 12(2): 38-43.

[11] Ito K., Aino S. Illustrations of Tree Disease and Insect Pests[M]. Tokyo: Sohbun, 1976.

[12] Jones H. G. Application of thermal imaging and infrared sensing in plant physiology and ecophysiology[J]. Adv. Bot. Res., 2004, 41: 107-163.

[13] Kozlowski T. T. (Ed.). Water Deficits and Plant Growth[M]. New York: Academic Press, 1976.

[14] Kozlowski T. T. Tree growth in response to environmental stresses[J]. J. Arboric., 1985, 11(4): 97-111.

[15] Liu G., Kennedy R., Greenshields D. L., Peng G., Forseille L., Selvaraj G. Detached and attached arabidopsis leaf assays reveal distinctive defense responses against hemibiotrophic colletotrichum SPP[J]. MPMI, 2007, 20 (10): 1308-1319.

[16] McElrone A. J., Sherald J. L., Forseth I. N. Effects of water stress on symptomatology and growth of parthenocissus quinquefolia infected by xylella fastidiosa[J]. Plant Dis., 2001, 85(11): 1160-1164.

[17] Mudd J. B., Kozlowski T. T. (Eds.). Responses of Plants to Air Pollution[M]. New York: Academic Press, 1975.

[18] Omasa K., Hashimoto Y., Aiga I. A quantitative analysis of the relationships between SO_2 or NO_2 sorption and their acute effects on plant leaves using image instrumentation[J]. Environ. Control Biol., 1981, 19(2): 59-67.

[19] Omasa K. Image instrumentation methods of plant analysis[J]. New Series, 1990, 11: 203-243.

[20] Omasa K., Saji H. S. Youssefian and N. Kondo (Eds.). Air Pollution and Plant Biotechnology-Prospects for Phytomonitoring and Phytoremediation[M]. Tokyo: Springer-Verlag, 1990, 343-359.

[21] Omasa K., Hosoi F., Konishi A. 3D lidar imaging for detecting and understanding plant responses and canopy structure[J]. J. Exp. Bot., 2007, 58: 881-898.

[22] Pearce R. S. Plant Freezing and Damage[J]. Annals of Botany, 2001, 87: 417-424.

[23] Pirone P. P. Disease and Pests of Ornamental Plants[M]. New York: John Wiley & Sons, 1978, 3-471.

[24] Richardson M. D., Karcher D. E., Purcell L. C. Quantifying turfgrass cover using digital image analysis [J]. Crop Sci., 2001, 41: 1884-1888.

[25] Schoeneweiss D. F. Infectious diseases of trees associated with water and freezing stress[J]. J. Arboric., 1981, 7: 13-18.

[26] Thorne E. T., Stevenson J. F., Rost T. L., Labavitch J. M., Matthews M. A. Pierce's disease symptoms: comparison with symptoms of water deficit and the impact of water deficits[J]. American Journal of Enology and Viticulture, 2006, 57, 1-11.

[27] Vollenweider P., Gunthardt-Goerg M. S. Erratum to "Diagnosis of abiotic and biotic stress factors using the visible symptoms in foliage"[J]. Environ. Pollut., 2006, 140: 562-571.

[28] Wang F., Yamamoto H., Iwaya K. Quantitative research on vigor of ginkgo trees hit by Typhoon 0613 with ground-based digital image analysis[J]. J. Nat. Disaster Sci., 2008a, 30(1): 45-53.

[29] Wang F., Yamamoto H., Ibaraki Y. Measuring leaf necrosis and chlorosis of bamboo induced by typhoon 0613 with RGB image analysis[J]. J. Forestry Res., 2008b, 19(3): 225-230.

[30] Wang F., Yamamoto H., Ibaraki Y. Responses of some landscape trees to the drought and high temperature event during 2006 and 2007 in Yamaguchi, Japan[J]. J. Forestry Res., 2009a, 20(3): 254-260.

[31] Wang F., Yamamoto H., Ibaraki Y. Transpiration surface reduction of kousa dogwood trees during seriously losing water balance[J]. J. Forestry Res., 2009b, 20(4): 337-342.

[32] Wang F., Yamamoto H., Ibaraki Y., Iwaya K., Takayama N. Evaluating ginkgo leaf necrosis and asymmetric crown discoloration induced by Typhoon 0613 with RGB image analysis[J]. J. Agri. Meteorol., 2009c, 45(1): 27-37.

[33] Wang F. Persistent and advanced reddening of sweetgum leaves after major veins severing[J]. Journal of Forestry Research, 2010, 21(4): 465-468.

[34] Wharton P. S., Julian A. M., Connell R. J. O. Ultrastructure of the Infection of Sorghum bicolor by Colletotrichum sublineolum[J]. Phytopathology, 2001, 91(2): 149-158.

[35] Wood B. W., Reilly C. C., Tedders W. L. Relative susceptibility of pecan cultivars to fungal leaf scorch and its interaction with irrigation[J]. Hortscience, 1995, 30(1): 83-85.

[36] Yapp R. H. Spiraea Ulmaria L. and its bearing on the problem of xeromorphy in marsh plants[J]. Ann. Bot., 1912, os-26: 815-870.

[37] Zhao J-b, Zhang Y-p, Tan Zh-h, Song Q-h, Liang N-sh, Yu L, Zhao J-f. Using digital cameras for comparative phenological monitoring in an evergreen broad-leaved forest and a seasonal rain forest[J]. Ecological Informatics, 2012, 10, 65-72.

(原文发表于 *Ecological Informatics*, 2012, 12: 16-22)

第五部分　松树枯萎

气象灾害环境下的松树枯萎*

王　斐

（山东省林业科学研究院，250014，济南）

1. 引言

松树枯萎病（也叫"松材线虫病"）在东亚地区较为流行，美国、墨西哥以及欧洲国家（如葡萄牙）的局部地区有时也有松树枯萎病的发生。许多国家已将该病列为重点防范的检疫对象。欧洲大陆的许多国家为了限制松材线虫的传入而开展了大量研究和防控，并且禁止从北美松材线虫分布区进口松木切片等。到目前为止，识别松树枯萎病的关键在于是否有松材线虫（Dropkin，Linit，1982；陈玉惠等，2001）的存在。对松树枯萎病的防控也主要集中在对松材线虫及其媒介昆虫的检疫和防治上。松树枯萎病与其他因素诱发的松树枯萎之间存在一些概念和统计学上的混淆，不同诱因造成松树枯萎的全球数据资料至今尚未见到。有关枯死松树中松材线虫检出状况的报道大多出现在早期的研究中，近期的研究报道中较为少见，这或许是因为其已经是确定无疑的事实。不过，在这些研究报道中，松材线虫的检出率不算很高。对中国国内十几个相关的研究报道进行统计后发现，从枯死松树上平均检出松材线虫的比率仅为 20.7%（汪来发等，2004；朋金和等，1993；钱大益等，1991；李刚等，1991；徐福元，1993；王兴周等，2004；王明旭等，2007；黄金水等，2009；杨希等，2006；黄任娥等，2008；张阜嘉，2009），大量松树枯萎是基于不明原因。在中国浙江和福建等省份的相关研究表明，枯死的松林中松墨天牛所携带的线虫种类繁杂，其中最多的是拟松材线虫（*B. mucronatus*），其次是松材线虫以及其他种类的线虫，也有相当一部分松树上没有发现线虫的寄生（Kishi，1995）。尽管在中国和日本已发现拟松材线虫可以使一些松树致病，但是在欧洲和北美大陆该线虫几乎没有致病能力（Wingfield，1983）。

尽管在北美大陆广泛分布着拟松材线虫及其媒介昆虫（Mamiya，1988），却尚未发生像东亚那样的松树枯萎现象（Wingfield 等，1984）。而且认为北美大陆的松树树种对该

* 基金资助：国家自然科学基金（31170671）。通讯作者：王斐，日本鸟取大学农学博士，山东省林业科学研究院研究员，从事树木生态生理学、灾害气象学和光谱分析等领域的研究。

病源线虫有抗性(Yang 等，2003)。然而，目前一些引进的北美松树树种在东亚(Shu 等，2006)和中国(Li，Wang，1997；Xiong，2009)同样也会感染松材线虫病(李修鹏等，1997；熊大斌等，2009)。

尽管日本制定了相关的法令来强制防控松材线虫(JNFPS，1997；Kishi，1995)，但在大量使用化学药剂、扑杀媒介昆虫、砍伐枯萎树等方法后，松树枯萎依然难以控制(Kishi，1995)。尽管在中国和韩国进行了大规模的检疫和防治，最近几十年松树枯萎病的发生仍然有增加的趋势。在美国，尽管松材线虫及其媒介昆虫分布广泛，但松材线虫病并未流行，只是被用做观赏的外来松树树种受害最为严重(Wingfield，1983)。

因为敏感寄主的死亡较为迅速，再加上需要媒介昆虫传播，所以松材线虫病的流行较为特殊。松材线虫或许可以因非实质原因而传播到已经枯萎或正在枯萎的松树植株上(Wingfield，1983)，而且据报道，无菌松材线虫对健康松树没有致病力(迟树友等，2006)。小田久五(1967)在对枯死松树的调查中发现，被害木上蛀干害虫的数目差异很大，甚至有些树木上没有蛀干害虫，其伐根也没有树脂流出，且断面干燥。松材线虫接种试验的薄弱环节在于日本和美国不同地点的接种结果相悖(Bain and Hosking，1988)。因此，松树枯萎病的流行学至今依然存在着不确定性。

就树木、寄生物和环境的三角关系来说，松树是一种特殊的树种，其对胁迫的独特适应和松脂的生产能力深受人们的喜爱(尤其是在东亚的日本、中国和韩国)，并且在全世界范围内被广泛引种栽培。在日本、中国和韩国甚至形成了独特的松树文化，并建起了众多松脂生产企业。松树也被认为是沿海防护林不可替代的树种，尽管在此环境中它们也经常遭受极端环境胁迫的袭扰。在经济效益的驱动下，松树往往被扩大栽培到不适合它们生长的环境中。在人工林内松树枯萎病比天然林内更加严重的事实表明，部分松树的栽培属于逆境栽植。松树枯萎病的发生或许与极端气象环境有关，也与这些逆境栽培的松树适应性有关。显然，环境胁迫或逆境条件可直接影响寄主的活力和抗性，并导致它们对松材线虫的敏感。

鉴于树木表形特征的复杂性，在正常天气条件下，气象因素的影响很难从其他因素的作用中分解出来。然而，极端气象事件往往成为树木生长的限制性因素、灾害来源，并造成对树木的伤害，即使是抗性较强的松树也不例外(Hitou，1964)。这些逆境栽培的松树往往会不断遭受极端环境的袭扰。经常发生的情况是在一次伤害尚未痊愈之前，环境胁迫便再次发生。遭受严重伤害之后，部分松树迅速枯萎，因为扰乱了树体的水分和能量平衡，甚至超出了其忍耐阈值极限。另一些尚未超过其忍耐极限的植株出现了衰退迹象。由于特殊的适应能力，与经常缩减其蒸腾表面积或回枯的树种不同，松树往往减少松脂生产(Lorio，Hodge，1968)，并发生整株枯萎。松树往往受害较迟，一旦受害又较为严重。许多松树枯萎的高峰可以追溯到几年前的极端气象事件(Wang，2012)，此类延迟往往成为诱因分析中误判的根源。

许多研究者认为松材线虫病的发病受环境条件的影响(Kishi，1995；徐克勤等，1996)。卢瑟福德(Rutherford，2006)认为温度是限制松材线虫病发病的重要条件，且在 7 月份平均气温低于 20 ℃的地区没有该病的发生。尽管如此，在 7 月平均气温高于 20 ℃的地区也只有全球局部地域发生了松树枯萎病。玛米亚(Mamiya，1984)的研究表明，干旱和高温环境可以加重病症的程度。也有不少研究得出了在降水偏多的年份少有病症发生的结论(Suzuki,1984),这意味着水分和能量失衡的影响相当于生物病原的作用,甚至更加重要。

在东亚地区,松树对极端气象事件的响应对松树枯萎起着至关重要的作用。通过不同地区的气象差异来比较松树枯萎的研究并不多见。因此,本研究试图通过比较研究东亚中、日、韩三国和美国的环境条件来探讨松树枯萎病发生和流行的可能原因。我们假设松树对环境的特殊适应性和灾害环境是东亚地区松树枯萎的重要激发因素。

2. 材料和方法

本文应用气象数据的对比研究了松树枯萎与夏季干旱和超强台风/飓风等气象极端事件的关系。数据来自东亚、北美、欧洲、地中海、澳大利亚、新西兰和南非等地区和国家,如表1所示:

表1　　气象数据聚类分析中的城市列表

地中海以外的地区		东亚		美国		地中海地区	
序列号	城市	序列号	城市	序列号	城市	序列号	城市
1	维也纳	11	大阪	9	波尔多	10	里斯本
2	明斯克	13	广州	12	札幌	23	洛杉矶
3	巴黎	14	重庆	19	伯明翰	47	卡那封
4	阿姆斯特丹	15	南京	20	新奥尔良	48	卡拉沙
5	基辅	16	上海	21	杰克逊	49	德班
6	布拉格	17	釜山	22	阿什维尔	50	伊斯坦布尔
7	伦敦	18	迈阿密	25	亚特兰大	52	雅典
8	柏林	33	仙台	26	巴尔的摩	54	罗马
40	布达佩斯	34	福冈	27	普林菲尔德	55	安曼
41	斯德哥尔摩	35	名古屋	28	圣路易斯	56	巴库
42	华沙	36	宁波	29	华盛顿特区	53	耶路撒冷
43	赫尔辛基	38	武汉	30	达拉斯沃斯堡		
44	布加勒斯特	39	厦门	31	史密斯堡		
45	吉朗	58	合肥	32	普罗维登斯		
46	堪培拉	59	杭州	51	萨拉热窝		
24	墨西哥城	60	东京	61	惠灵顿		
57	卡潘	37	青岛				

2.1 一些典型城市气候数据的聚类分析

本文对全球所研究地区的上述61个典型城市的各月降水量、平均日最低气温和最高气温的常年值进行了欧式平方距离的聚类分析研究。

2.2 气象数据、松树分布以及松树枯萎病的地理统计分析

1971～2010年这40年间，日本各行政县之降水量数据的平均值由日本气象厅的全自动气象观测系统之主要站点数据统计而来，而美国各州(夏威夷和阿拉斯加除外)的降水量数据来自于美国Palmer干旱研究中心。在此基础上，应用ARCGIS 9.3系统的径向基函数地理统计分析对日本和美国夏季8月降水量的年间变动性和年内各月间变动性的描述统计参数(标准误差值)进行了分析，并分别绘制了日本和美国的40年降水量年间变动等值线图和年内各月间降水变动的等值线图。

日本各地遭台风袭击时风速达1～2级的台风数据资料同样来自日本气象厅的全自动气象观测系统，相应的美国飓风数据则来自美国国家飓风中心。以此为基础，应用ARCGIS 9.3系统制作出1～2级台风/飓风袭击回归年的区域符号图。

在分析和研究发生在日本和美国的典型台风/飓风数据的基础上，绘制了典型台风(如台风9119号)/飓风袭击时的降水空间分配特征图和瞬间阵风风速图。

在收集大量松树分布情况矢量图的基础上，应用ARCGIS 9.3系统绘制了世界范围内松树潜在分布区和引种栽培区域图以及松材线虫病发生点分布图等。

东亚三国松树枯萎病的点密度分布分析图以及各大洋区域热带风暴点密度分布图等同样是运用ARCGIS 9.3系统、采用核函数密度法完成的。世界范围内以及局部海域最大阵风风速超过113节的超强台风/飓风之时段(台风路径上每日0、6、12、18时四个时间点风速不低于64节)的点分布图的制作是利用台风(飓风)的路径地理坐标和风速资料，应用ARCGIS 9.3系统的Kernel点密度分析功能制作完成的。

2.3 相关指数的构建

文中涉及的相关统计指数计算如下：

2.3.1 降水波动指数

该指数综合反映了日本和美国各地1971～2010年这40年间以及1年内12个月间的降水变异特征，算法见公式(1)，并用这些数据利用ARCGIS 9.3系统的径向基地统计分析功能绘制了分析图：

$$\text{V-index}x_k = \left[\sqrt{\sum_{i=0}^{39}(x_i - \bar{x})^2/1560} + \sqrt{\sum_{j=0}^{11}(y_j - \bar{y})^2/132}\right]/2 \tag{1}$$

式中，$\text{V-index}x_k$为第k州(美国)或县(日本)的降水波动指数，x_i为该州/县最近40年中第i年的8月份降水量，$\bar{x}$为这40年中8月份降水量的平均值，y_j为该州/县第j月份的降水常年值，$\bar{y}$为年内各月降水量常年值的平均值。

2.3.2 飓风干旱关联指数

该指数通过美国 12 个无飓风登陆的年份之 8、9、10 月平均帕尔默(Palmer)干旱指数与 10 个登陆飓风最多的年份之平均 Palmer 干旱指数的差值来反映登陆飓风与干旱的关联程度,具体算法见公式(2),并应用 ARCGIS 9.3 的地统计分析功能绘制了分析图。

$$\text{HuP-index}_k = ((\sum_{i=1}^{12} \text{PHDI}_i)/12) - ((\sum_{i=1}^{10} \text{PHDI}_j)/10) \quad (2)$$

式中,HuP-index_k 为美国第 k 州的飓风干旱关联指数,PHDI_i 为该州 1895 年以来 12 个无飓风登陆年份在第 i 年($i=1,2,\cdots,12$)之 8、9、10 月份的 Palmer 干旱指数值,PHDI_j 为该州 10 个飓风登陆最多年份在第 j 年($j=1,2,\cdots,10$)之 8、9、10 月份的 Palmer 干旱指数值。

3. 结果和分析

3.1 松属树种的分布和松树枯萎病发生的格局

全世界有 100 多种松树,分布在北美大陆、欧亚大陆和中美、北非山地等区域。在南半球的新西兰、澳大利亚、智利和南非等地也有引种栽培(见图 1a)。在高纬度地区常见少数几种松树占据着广阔的地域范围;而在低纬度地区,松树常分布在高海拔地区且树种数量多,但局限在狭窄(适生)的特殊地域范围内。因地理隔离和气候环境差异,南半球没有松树的天然分布。而在新西兰、智利、南非等国比较湿凉的温带环境中,引种的辐射松等树种生长较为成功,其中在新西兰的栽培区地处南纬 38 度以南的地域(见图 1b)。

除此之外,在北纬 25～35 度,气候受副热带高压的影响而多有荒漠和半荒漠的大陆气候区域。在这些地区,除了有少量的高山松树树种资源外,基本上没有松树的分布(见图 1c),如非洲的撒哈拉沙漠,西亚、中亚和北美的荒漠地带。但是,在东亚沿海地带,特殊的季风气候和热带气旋带来的大量降水使该地域多有松树的分布和栽培,甚至成为沿海岛屿和海岸林的主栽树种。受飓风降水的影响,北美东海岸一带同样有为数不少的松树分布和栽培。

松树枯萎病流行于东亚的中、日、韩地区(见图 1d),局部发生在美国、墨西哥和葡萄牙等国(见图 1e)。在流行的地区,大多发生在中低纬度和低海拔地区,高纬度、高海拔地区少有发生。巧合的是,发病和流行的松树树种几乎都是那些分布和栽培范围广、在低海拔地域广为栽植的松树树种,如中国的马尾松、日本的黑松和赤松、北美的湿地松、欧洲的赤松等。在台风多发的沿海地带也常有松树枯萎病的发生,且存在一些松树枯萎病的集中爆发地域。在东亚的中、日、韩三国,松树枯萎病大多从东南沿海向西北内陆方向扩展蔓延的特征也表明,西北太平洋热带风暴对这一地区的松树枯萎存在影响和作用。在唯一发现有松树枯萎病的欧洲国家葡萄牙,该病发生在靠近里斯本和塞巴图尔的特茹河和萨多河沿岸地带人工海岸松栽培区(见图 1f),这一地区位于北纬 37～38 度,为欧洲松树树种分布的南界地域且具有典型的地中海气候特征。在美国,松树枯萎病主要发生在引进的观赏用松树树种，如欧洲赤松、欧洲黑松、日本赤松和日本黑松中，美国本地松树树种只有少数局部地域出现了枯萎病症。松材线虫及其媒介昆虫在北美大陆分布广泛，在未传染松材线虫病的欧洲大部分地区也有媒介昆虫种类和拟松材线虫的分布。在南半球松树引种栽培区，目前也没有松树枯萎病发生和流行的趋势。

图 1　a 为世界松树分布和引种栽培地区轮廓图和松材线虫发生地域图；引种栽培松树的澳大利亚、智利、南非；b 为新西兰松树栽培区；c 为北纬 25～35 度没有松树自然分布的地区；d 为日本、中国、韩国松树枯萎病流行区；e 为美国和墨西哥松树枯萎病局部发生区；f 为葡萄牙地中海气候的松树枯萎病发生区(参见彩色附图 13)。

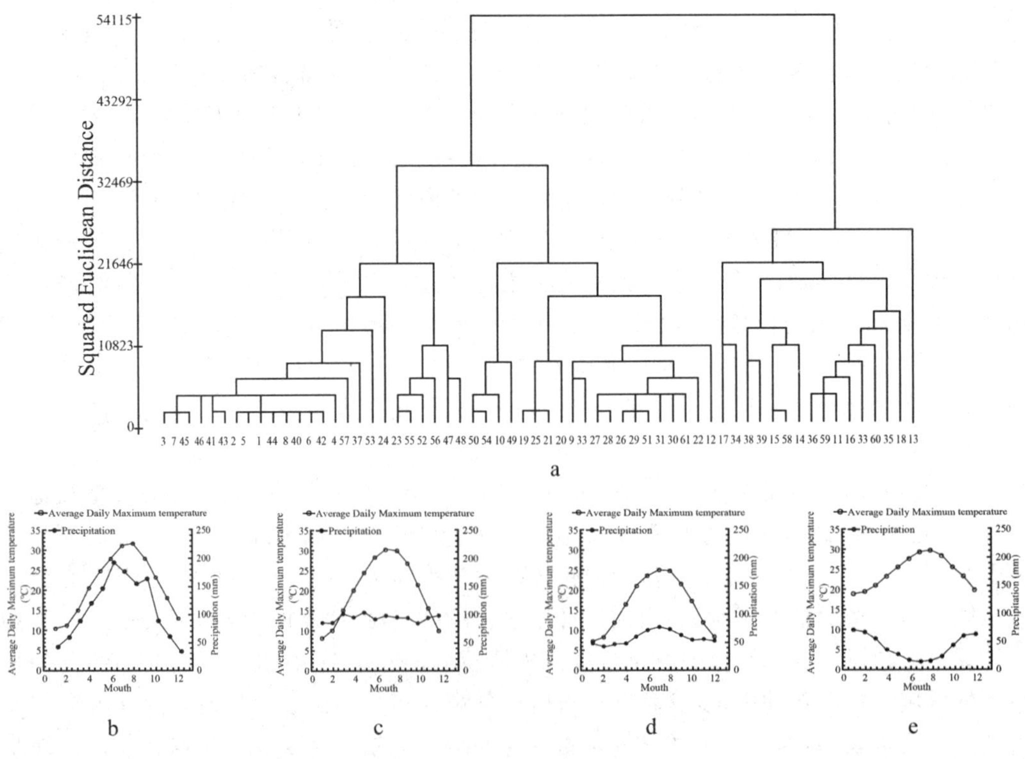

图 2　图 a 为世界范围内地处松树枯萎病流行区域不同距离的主要城市之气象要素的系统聚类分析结果；图 b 为东亚典型的夏季高温、集中降雨气候类型图；图 c 为美国大陆性气候类型图；图 d 为欧洲大陆地区类型图；图 e 为地中海气候类型图。

3.2 东亚、美国和欧洲气候的差异和比较

松材线虫及其媒介昆虫广泛分布于东亚和北美且受台风/飓风的袭扰。或许有人会问：为什么北美没有大面积流行松材线虫病？可能的原因之一是二者之间气象环境的差异。聚类结果呈现松材线虫病流行区、非流行区以及未发现区域之间的气候存在差异(见图2a)。

地中海以外之欧洲大陆的气候较为温和，变化幅度较小，所以采用欧式距离法聚类时，首先聚类的是一些欧洲国家的主要城市，然后是美国的一些主要城市，最后才是东亚的中、日、韩三国的城市(见表1)。鉴于个别边远和高海拔的城市具有较为独特的气候特征，往往会出现例外情况，因此将它们在一起归并。如日本北海道的札幌市常与美国城市归并在一起，美国加利福尼亚州、澳大利亚的卡纳封和卡拉沙等常与葡萄牙里的里斯本等地中海城市归并在一起，法国的波尔多常与美国罗得岛州的普罗维登斯归并在一起。相反，美国的迈阿密常归并到东亚中、日、韩三国的范围之内。

尽管选用的城市组合不同其分类结果并非一成不变，但大的地域性特征和变化趋势基本是保持稳定的。各气候区域的特点则突出地反映在平均气候特征图之中(见图2-b、c、d、e)，在东亚松树枯萎病流行的气候区夏季气温明显偏高，且存在一个夏季“V”形的降水间歇期。而欧洲地中海以外区域的夏季气温温和且年内降水分布均匀。实际上，单独从维度的分布上来看，欧洲整体上位于北纬40度以北的地域，美国东部地区基本分布在北纬30～45度，而东亚除了日本本岛和朝鲜半岛以外，维度可南伸到北纬20度。与东亚相比，美国东部地区的大陆性气候类型更接近欧洲气候区，而东亚的气候类型则属于典型的季风性气候。

通过对1970～2010年日本46个都、道、府、县(见图3c，不包括冲绳)和美国48个州(见图3d，不包括夏威夷和阿拉斯加)以及中国27省市的逐月降水量进行变异分析，结果表明日本和中国各地各月之间降水的变动性远大于美国各州，且在日本和美国均存在地域间差异。在东亚，中国和日本的降水分布不均、波动性大，雨季降水较为集中，甚至能占全年降水量的60%，个别地方可达70%～80%。这种降水特征使大量的雨季降水以径流形式直接排入大海，而且常伴随着地质灾害的发生，水资源的利用效率偏低。美国和欧洲的许多地区尽管绝对降水量不算太高，但相对均衡的降水分布特性使这些地域集中径流量相对较少，水资源利用效率更高些。

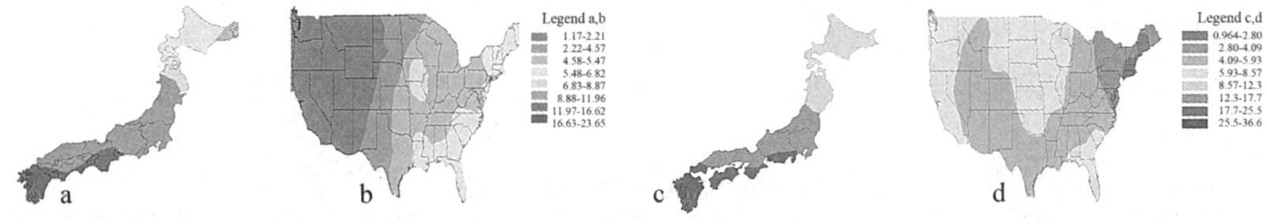

图3 1971～2010年日本和美国年度间与年内降水量标准误差的地理分布图。图a为日本8月份降水量标准误差的分布图(不包括冲绳)；图b为美国8月份降水量标准误差的分布图(不包括夏威夷和阿拉斯加)；图c为日本各月份降水量标准误差的分布图(不包括冲绳)；图d为美国各月份降水量标准误差的分布图(不包括夏威夷和阿拉斯加)。

在东亚地区，总的降水趋势是东南沿海大于西北地区，似乎与东南区域热带风暴袭击较频繁有关。与此同时，7、8、9 月份降水量的年度间变异存在明显的不同，且有日本各都、道、府、县大于美国各州的迹象(见图 3a、图 3b)。也就是说，地处东亚季风气候区和遭受较强台风袭击的日本存在着较大的年度间和年内各月间的降水波动特征，这意味着日本以及与其有相似的季风气候特征的中国和韩国面临着更大的春夏同旱、夏秋连旱的极端气候条件，尤其是其东南沿海地域。若与欧洲大陆区的国家进行比较，这种特点就更加突出，因为这些欧洲国家的气候波动性更小。在日本，北海道的降水波动最小，而且与美国东部地区相似——两地均未遭受松树枯萎病的严重危害似乎成了极端气象灾害事件导致松树枯萎病发生和流行的又一支点。

在近 30 年中，极端少雨的年份里，夏季最热的 8 月份日本有许多气象观测站点只有检测到几毫米的降水，这在常遭飓风袭击的美国东部地区较为少见。再加上东亚地区遭受台风袭击的概率本来就高，所以在这些地方发生少雨台风的可能性较大，尤其是台风发生最多的 8～9 月份。另外，在日本，8 月份降水量的标准误差大的地区松材线虫发生较早且严重，这说明夏季干旱高温和台风雨缓解干旱的能力是限制松树枯萎病发生和流行的重要因素。

在美国中西部地区存在一个年内各月间降水量标准误差相对较高的区域，这似乎与该地首先发现欧洲赤松等树种发生松材线虫病的事件吻合。在日本松材线虫病的主要发生区域内，8 月份降水量的标准误差则更高(见图 3c、图 3d)。而北海道等地至今尚未流行松材线虫病似乎与该地同美国大陆的降水等气候特征相似有关。不管怎样，降水量标准误差分布的差异足以说明日本和美国的气象环境存在着较大的区别。对美国而言，在降水很少的西部地区降水的波动性小，这些地区松树树种大多分布在比较湿润和低温的高山环境条件中，一些严重干旱的荒漠地区几乎没有松树的分布。

3.3 台风/飓风灾害频发地域的比较

尽管美国也会遭受来自北大西洋的飓风的袭击并深受其害，但与东亚的台风危害仍然有所差别。首先，在全球范围内，1995～2010 年的统计数据表明，各大洋海域发生热带气旋(风速大于 34 节)的频度不尽一致。发生次数最多的是影响东亚的中、日、韩三国最为严重的西北太平洋地区，平均每年发生 26 次以上，影响美国最多的北大西洋海域形成热带气旋的频率较低，平均每年只有 13 次(见表 2)，而飓风发生的次数会更少。这使得美国受飓风影响的概率小于东亚地区，再加上大西洋海水的相对低温，使飓风的强度和影响范围也受到了限制。此外，东北太平洋一侧以西行路径为主的飓风很少直接登陆美国本土。

表 2　　相关地区热带气旋与台风/飓风的发生频率

	南太平洋和印度洋	北西太平洋	北东太平洋	北大西洋
热带气旋次数	8.6	26.4	13.5	13.8
每年平均台风飓风次数	5.1	15.8	6.7	7.4

因此，每年登陆中国、日本的台风（风速不低于64节）和登陆美国的飓风（风速不低于64节）数各不相同。最多的是中国，每年约登陆3.29次；日本为2.86次；而美国只有1.79次（见表3）。在登陆美国的飓风中，发生在9月的最多，其次为8月和10月，再次是7月。而在登陆日本的台风中，8月发生的最多，其次是9月和7月。在中国，7、8、9月登陆的台风较为均衡。东亚地区的季风气候、梅雨的年季变化和"入梅""出梅"的早晚差异往往会导致局部地区的伏旱天气，这显然增加了台风与夏季干旱相关的可能性。尽管台风所携带的雨水可减缓夏季干旱的危害，但台风雨分布的不均匀性反而会加重局部地区的受害程度。此外，1995年以来在登陆的强台风和强飓风（中心气压低于975 hPa）中，9月10日前登陆者占日本登陆台风的60%之多，而在美国只占37.5%；9月15日后登陆美国的强飓风占50%，而同期在日本登陆的强台风只占35.7%。显然，相对于日本而言，有近2/3的美国飓风发生在9月10日以后，也就是说美国同时遭受飓风和夏季极端干热胁迫等灾害链袭击的概率低于日本等东亚国家（见表4）。此外，美国6、7、8月份发生较早的飓风主要集中在得克萨斯州和路易斯安那州，在美国东海岸和佛罗里达等州松树主产区飓风大多发生在9月之后。因此，美国松树主产区同时遭受飓风和夏季干旱灾害的概率比东亚地区也低得多。

表3　登陆美国、日本和中国的飓风（台风）的按月分布情况对比

美国大西洋海域 1851～2006	日本 1951～2009			中国 1949～1993			美国		
月份	总数	平均	（%）	总数	平均	（%）	总数	平均	（%）
1～4	0	0	0	1	0.016923	0.5917	0	0	0
5	0	0	0	2	0.033846	1.1834	3	0.080631	0.02027
6	19	0.12	6.7039	10	0.169231	5.9172	13	0.349399	0.087838
7	25	0.16	8.9385	27	0.456923	15.9763	36	0.967568	0.243243
8	74	0.48	26.8156	61	1.032308	36.0947	36	0.967568	0.243243
9	105	0.69	38.5475	53	0.896923	31.3609	43	1.155706	0.290541
10	51	0.33	18.4358	14	0.236923	8.284	13	0.349399	0.087838
11	5	0.05	2.7933	1	0.016923	0.5917	4	0.107508	0.027027
12	0	0	0	0	0	0	0	0	0
合计	279	1.79	1	169	2.86	1	148	3.29	1

表4　9月中旬前后日本和美国的强台/飓风（小于97.5 kPa）发生率的差别

	9月10日前	9月10日后	9月15日前	9月15日后
美国	37.5%	62.5%	50.0%	50.0%
日本	60.7%	39.3%	64.3%	35.7%

注：鉴于美国太平洋海域的飓风极少登陆，因此表中美国的飓风为大西洋海域的数据。

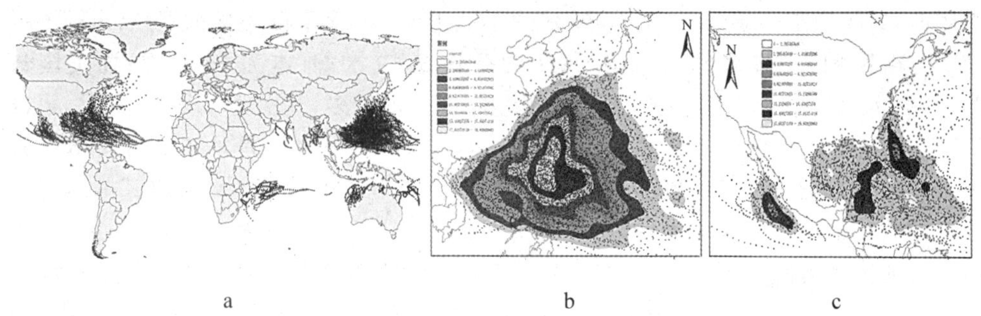

a　　b　　c

图 4　图 a 为世界范围内最大阵风风速超过 113 节的超强台风/飓风(每日 0、6、12、18 时中风速大于等于 64 节的时段)的点分布图；图 b 为西北太平洋区域该类数据的 ARC-GIS 9.3 系统 Kernel 点密度分布图；图 c 为北大西洋区域此类数据的 ARCGIS 9.3 系统 Kernel 点密度分布图(参见彩色附图 14)。

日本和美国遭受台风/飓风袭击的差异不仅表现在袭击的频率上，而且台风/飓风袭击的强度也不尽一致。最大阵风风速超过 113 节的超强台风/飓风时段(风速大于等于 64 节)的分布点数在影响东亚三国的西北太平洋区域非常密集(见图 4a)，尽管影响美国的北大西洋飓风区域之点数比其他大洋区域更密集，但与西北太平洋相比则微不足道。这种差异可在 ARCGIS 9.3 系统的 Kernel 点密度分布图上明显地反映出来(见图 4b、4c)。就日本而言，这种超强台风之台风时段的点密度分布与其年度间和年内的降水变动性的空间分布似乎存在某种相似的变化趋势和关联性(见图 4b，图 3a、3b，图 3c、3d)。这种特征同样在中国东南沿海和美国的东南局部地域有所表现。显然，这来自台风/飓风的补充降水效应的波动性。此外，由于台风/飓风登陆后强度的剧减效应，此类超强台风/飓风的风域只波及具有巨大版图的中国和美国之沿海地带，而对于版图狭长的日本则影响范围更加广泛，甚至可波及包括北海道在内的日本全境(见图 4b)，这意味着即使地处北纬 40 度以北的北海道地区也时常会遭受台风的袭击。

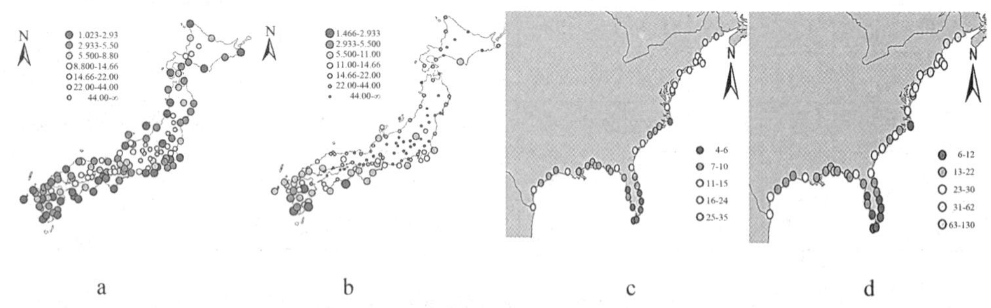

a　　b　　c　　d

图 5　日本各地和美国大西洋地域最大阵风风速超过 33 m/s(一级)和超过 42 m/s(二级)的台风(飓风)回归年值的比较。图 a 为最大阵风风速超过 42 m/s 的台风回归年值；图 b 为最大阵风风速超过 33 m/s 的回归年值；图 c 为 1 级以上飓风回归年；图 d 为 2 级以上飓风回归年(参见彩色附图 15)。

台风袭击日本各气象站点时风速达 1 级(阵风 33 m/s,见图 5a)和 2 级(42 m/s,见图 5b)以上的台风之回归年数最低值为 1～3 年,也就是说在个别地方几乎每年都有此类台风的袭击。尽管日本各地区遭受台风袭击的频率不同,但几乎各地均有台风风速达 1 级以上的经历。风速 2 级以上的台风袭击频率则存在明显的地域差异:位于九州东南的鹿儿岛、熊本等县属台风回归年最短的地区,其次是九州其他地域以及四国、中国地方和近畿等地区的一部分。最弱的地域当属北海道、东北地方以及日本海沿岸的县市等。显然,这与西北太平洋热带风暴来袭的方向相吻合。广阔的版图使美国遭受北大西洋飓风袭击的地域主要集中在东、南两侧,且最为严重的受害地是佛罗里达州,其次为得克萨斯州和北卡莱罗纳州。据美国国家飓风中心的报道,平均 1 级飓风(风速大于 33 m/s)最小回归年需要 4～6 年(见图 5c),2 级以上飓风(风速大于 42 m/s)最小回归年需要 6～12 年(见图 5d)。因此,可以说日本遭受台风袭击的广度和强度均大于美国。

在北半球,台风或飓风的共同特点是逆时针自转和顺时针的运行轨迹,这使得美国东部地区遭受飓风直击的概率减小,许多飓风在到达美国东海岸之前便转向东北的大西洋水域了。在能够到达美国东海岸的飓风中,相当一部分在到达时已经在很大程度上被削弱了,也有一部分仅仅以相对偏弱的左半侧掠过东部海岸地区。考虑到飓风强弱的空间分布非对称性,贾瑞尔(Jarrell,1992)在定义美国遭受飓风直击的范围时,将飓风眼左半侧等于半径和右侧 2 倍于半径的范围确定为台风直击区域。袭击美国南部地区的飓风同样具有这种非对称性特征,且常在强度更大的右半侧降水更加丰富。再加上阿巴拉契亚山脉造成的右半侧地形雨,尤其是南撤副热带高压脊与飓风交汇的协同作用使得这一区域的降水丰沛,以至于有人常以飓风移动速度来预测总降水量。在中国南部的广东省,尽管年降水量较高,但为数较多的西行台风路径使得广东位于强势的台风右半圆区,再加上极高的台风袭击频度和密度以及较低的纬度等,使该地台风灾害较为严重。从墨西哥湾进入美国南部地区的飓风登陆后风速常剧减成热带风暴或低气压,且由于这一地域的纬度接近北纬 30 度而受西风急流的影响,风在登陆后迅速右折。这与发生在日本地域的台风,尤其是从日本海受西风急流影响而加速北上之台风的持续强势有明显的区别。像台风 9119 号和 9918 号那样降水偏少的超强台风(见图 6)在北美地区也非常少见,这使得本来就比东亚地区遭受飓风袭击更少的美国东部地区面临无(少)雨飓风袭击的可能性更小了,即使在干旱到湿润的过渡地带(如美国得克萨斯州)也有同样的趋势。美国得克萨斯州刚好处在美国东部到西部降水量由高到低的过渡区,这与中国的南京地区相似。然而,两地降水梯度变换的方向正相反,中国的降水梯度是从左向右(由南向北)递减,而美国的降水梯度是从右向左(从东向西)递减。在中国,台风直击时的右半圆刚好处在降水量较低的地区,而且其运动轨迹也朝着降水量少的低平开阔区前进,因此发生少雨强风的概率大。相反,在得克萨斯州(或与路易斯安那州交界处)登陆飓风的右半圆刚好面临降水较多的湿润地区,且运行轨迹也朝降水分布最高并有山地阻截的地区运行,从而导致呈现少雨飓风的可能性小。从地理位置上看,中国东部沿海与美国东海岸遭受台风/飓风袭击的情况更相似,但大地形的差异又使得二者受害程度明显的不同。就美国东海岸而言,飓风路径大多呈弧状,从沿海岸擦边而过(见图 7a),而在中国浙江沿海有直击大陆的趋势,且台风时段的阵风往内陆渗透的幅度比美国东海岸更加广泛(见图 7b)。二者相比,袭击中国浙

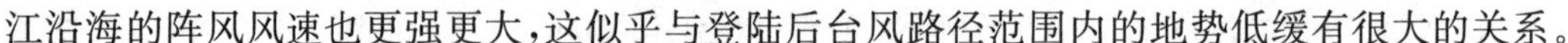

江沿海的阵风风速也更强更大，这似乎与登陆后台风路径范围内的地势低缓有很大的关系。

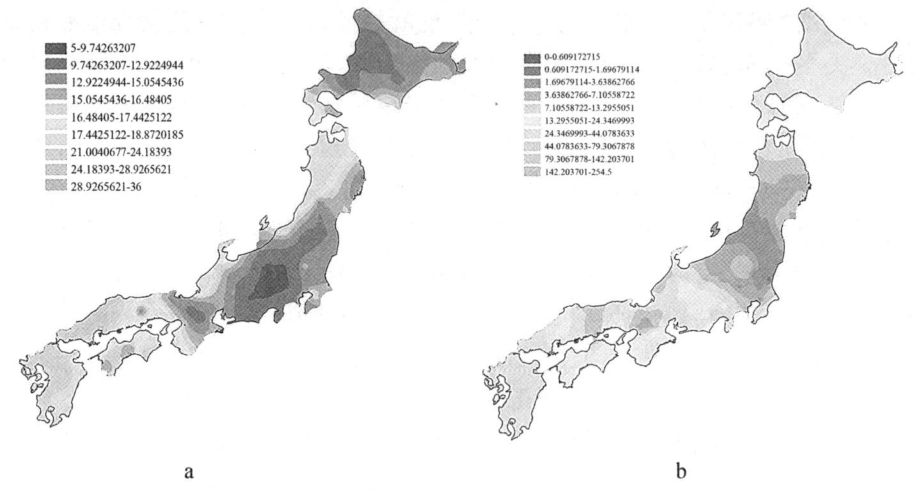

图 6　超强台风 9119 号袭击日本期间的最大阵风风速(a)和降水分布图(b)。

相比之下，日本遭受台风袭击的特征有更加鲜明的特点。作为东亚太平洋岛国，日本的版图狭长，一些大型台风甚至能横扫日本本岛。鉴于日本版图的“弓”形特征，有两种路径的台风对日本影响大，造成的危害也更严重：一种是从九州西南侧登陆的台风，影响范围纵跨九州、中国地方，直到本州中部，最典型的有 1999 年的台风 18 号、1991 年的台风 19 号和 2004 年的台风 18 号；另一类是从四国—近畿登陆，影响范围纵贯本州中部和北部直到东北地区的强台风。历史上，对日本影响最为深刻的两次台风——室户台风和伊势湾台风就属于此类台风。

台风/飓风和夏季干旱等气象因素构成的灾害链对作物和树木的伤害更加严重。然而，美国南部地区在发生为数不多的连续干旱时，几乎很少有飓风的袭扰。大西洋副热带高压的疲弱使许多热带气旋到达北美大陆之前就转向大西洋，例如 1999～2001 年美国南部地区发生的大范围干旱，这三年仅有 4 次飓风登陆美国，其中多数只在美国擦边而过(掠过佛罗里达半岛)，而且强度弱、影响范围小。在这之后的 2004 年和 2005 年两年内有 10 次飓风登陆，其中包括一些引发狂风暴雨和严重风暴潮的飓风事件。在为数众多的飓风袭击时，发生干旱的概率非常低。这种趋势在无飓风登陆和飓风密集登陆年份之 8、9、10 月份的飓风干旱关联指数分布状况中有充分的体现(见图 7c)，尤其是在东南沿海和墨西哥湾一带。这主要源于飓风带来的补充水分。不管是飓风登陆少导致干旱的发生，还是干旱年份飓风不具备登陆的条件，其结果都是飓风登陆和干旱同时发生的概率减小。尽管日本和中国等东亚国家同样在干旱年份登陆台风数有所减少，但由于台风发生次数本来就较多且有西行台风等的影响，故该地域干旱年份出现极端气象灾害链的可能性相对较大，最典型的事例就是日本松材线虫病极端高峰年的 1978 年，大范围的持续干旱和台风 7808 号等的肆虐诱发了日本松树的大量枯死。1978 年夏季 7、8、9 月份的干旱和 8 月初袭击日本的台风造成东京等众多地区出现了松树枯萎的首次高峰峰值。在这种环境下，台风过境时没有或几乎没有降水且伴随着接近 30 m/s 的强风和超过 30 ℃的高温，台风过境后又出现了持续的高温少雨。1978 年台风 7808 号袭击东京后十几天内无降水，并且 40 天内仅有 20 mm 的降水。

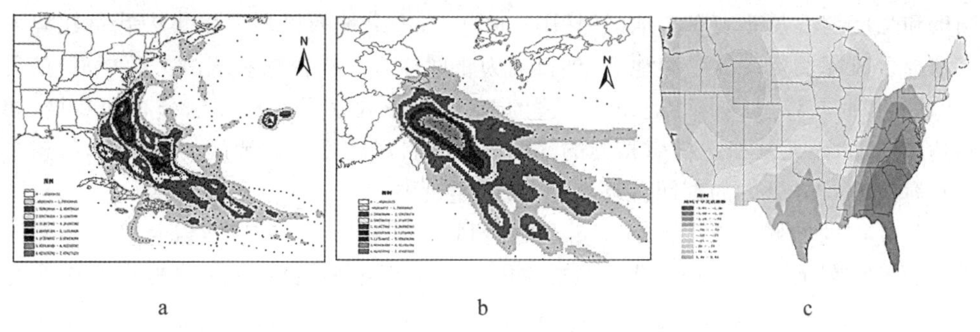

a　　　　b　　　　c

图 7　美国东海岸(a)和中国浙江沿海(b)登陆的强台风/飓风之时段(每日 0、6、12、18 时风速不低于 64 节的时段)的点密度分布图；美国在无飓风登陆和飓风密集登陆年份之飓风干旱关联指数的克里格地统计分析图(c)。

1991 年，日本冈山县 8、9 月份的降水量(113.5 mm)仅为常年的 45.0%，9 月 27 日袭击日本大多地方的台风 9119 号在冈山县测得最大阵风风速为 39.4 m/s，10 分钟平均最大风速为 19.5 m/s，这使其成为冈山县有详细记载以来的第二强台风。这场台风袭击冈山县期间的日降水量仅为 1 mm，最高、最低气温分别为 31.7 ℃和 22.4 ℃。台风袭击之后数日内无补充降水，10 月份的降水量也只有 40 mm，是常年的 45.7%，而 10～12 月份的降水量也只有 142 mm，是常年降水量的 77.2%。冈山县当年松树枯损量达 $11.5\times10^4\ m^3$，这是冈山县第三次松树枯萎的高峰值。显然，台风蕴含的巨大能量既可以带来可观的降水，但如果风速过大、过强而造成巨大的海潮，则可以淹没大片农田和林地，甚至会造成高大树木的机械损害。这在台风登陆的沿海地带尤为突出。而在远离登陆地点和海岸线区域，尽管风速的降低会减少造成损害的概率，但随着距离的增加可导致降水的减少，这会使遭遇少雨强风天气的可能性加大。这种灾害的相关性往往会造成更大的危害。

3.4　松树枯萎病与不同地理环境的关联分析

综上所述，美国大陆东部松树集中分布和栽培区的气候与东亚的日本、中国和韩国松树分布栽培区既有常受台风/飓风袭击、降水丰富等相似之处，也存在明显的区别。相对于日本、中国东部和韩国，美国东部地区降水等气象要素的波动性相对较小，飓风袭击的频度和强度稍弱，袭击时的强风常伴随着丰富的降水，且飓风袭击期间的气温偏低，等等。这使得该地气候对松树枯萎病的诱发作用远不及东亚三国。因此，不难理解为何至今美国尚未发生大面积的松材线虫病的流行，尤其是适应本地环境的美国乡土松树树种。

在世界主要大陆之间，欧洲遭遇台风/飓风袭击的概率最小(见图 4a)。到目前为止，欧洲是松材线虫病罕见且未流行的地区，这也是研究人员倾力探讨的问题之一。尽管在欧洲许多国家已经发现有拟松材线虫的存在，甚至在个别国家也有松材线虫的存在，而且有大量研究表明，欧洲赤松是感病的树种，但该病的大发生几乎没有报道。在葡萄牙里斯本附近发生的松树枯萎病表明，尽管该地遭受飓风袭击的概率很小，但地中海型夏季干热的气候加上局地的干热风袭击也可以诱发松树枯萎病的发生。况且，葡萄牙也地处大西洋热带风暴偶尔光顾的区域。在同样遭受台风袭击且气候干旱的澳大利亚地区，台风发生频度和强度均不及北太平洋地区（见图 4a），且该地引种的松树大多集中栽培在澳大利亚东南部台风鞭长莫及的湿润地区，到目前为止同样未见松树枯萎病的大面积流行（见图 1）。除此之外，在许多热带气旋经常出没的地区，由于几乎没有松树的分布和栽培（如非洲马达加斯加）而无从谈起松树枯萎的问题。地处台风密度最大地区的菲律宾，松树主要分布在海拔 800 m 处的局部阴湿环境中，且数量有限。

正是美国的气候与东亚地区均遭受强台风/飓风袭击的相似性，以及干旱环境的存在等，才使得美国的一些观赏用引进松树树种和个别本地低海拔分布的松树树种在一些风口、干旱和半干旱的环境地区（如伊利诺斯州、密苏里州、得克萨斯州、佛罗里达州、马里兰州及特拉华州）发生过松材线虫病的病例。从图 8 可以看出，在美国，松树天然分布较少的中西部地区栽培的外来松树树种枯萎范围较广（见图 8c），这些地方也是美国降水波动系数较大的地域（见图 8b）。美国本地松树树种没有该病的流行似乎也与这种波动性偏低有关。在日本，松树枯萎病的流行同样大多发生在东北地区以南降水波动系数较大的地区（见图 8a）。除此之外，美国大西洋飓风袭击的风口地带，如密西西比河河口、切萨皮克湾沿岸等也有发生，这与中国长江和珠江口等松树枯萎病集中爆发地有相似之处（见图 1）。

图 8　图 a、图 b 分别为用日本和美国各县/州的降水波动指数通过 ARCGIS 9.3 系统的径向基函数进行的地统计分析结果图；图 c 为北美地区松树分布或潜在分布特征图及枯萎松树中发现松材线虫的样点（参见彩色附图 16）。

美国中西部地区外来松树树种枯萎病的发生似乎还与该地常遭受干旱袭击有关。美国中西部地区常发生极端的干旱事件。位于美国中西部地区的伊利诺斯州地处湿润地区向干旱地区的过渡地带。不像美国东南部各州，该州基本地处东和南两面飓风降水的“雨影”区，且有时也受飓风的袭扰。地处落叶阔叶区的伊利诺斯州松树资源较为贫乏，以观赏树种栽培的欧洲赤松、欧洲黑松、日本赤松和日本黑松 1980～1981 年在该州 102 个县中的 50 个县发生枯萎，而地处同一地域的美国本地松树则无发病的迹象。正常情况下，欧洲赤松主要分布在北纬 45 度以北的地带，将该树种引种到依然有松材线虫病流行的北纬 37～38 度显然超出了其适生范围。对该州的气象数据分析表明，1980 年为中等水平的气象干旱年，在此之前的 5 年累算降水量却是严重的干旱周期年（见图 9a）。在长期气象干旱的条件下，1980 年干旱的持续诱发了观赏用欧洲赤松松材线虫病的发生。从图 9b 可以看出，松树枯萎存在滞后的现象，尤其是 1980 年仍然表现出了滞后于降水的松树枯萎峰值，这也说明要致死该州较大径级的松树的确需要一个过程。1980 年，由于区域性干旱，同样导致邻近的密苏里、阿肯色等美国中西部若干州的观赏松树树种发生了松树枯萎病。

类似的松树枯萎滞后现象在中国浙江省舟山市定海区表现得更加突出，具体表现在 5 年平均降水量和松树枯萎数量的变化趋势之中（见图 9c）；这种滞后还体现在 1997 年中国松脂主产区农作物干旱成灾率与 1998 年中国松脂主产区松脂产量之间紧密的反函数相关关系（见图 9d）之中。不仅如此，我们甚至发现了松树枯萎与强风和干旱事件的显著相关性，而且松树流脂的能力甚至被用作早期判别松树枯萎病是否发生的指标。

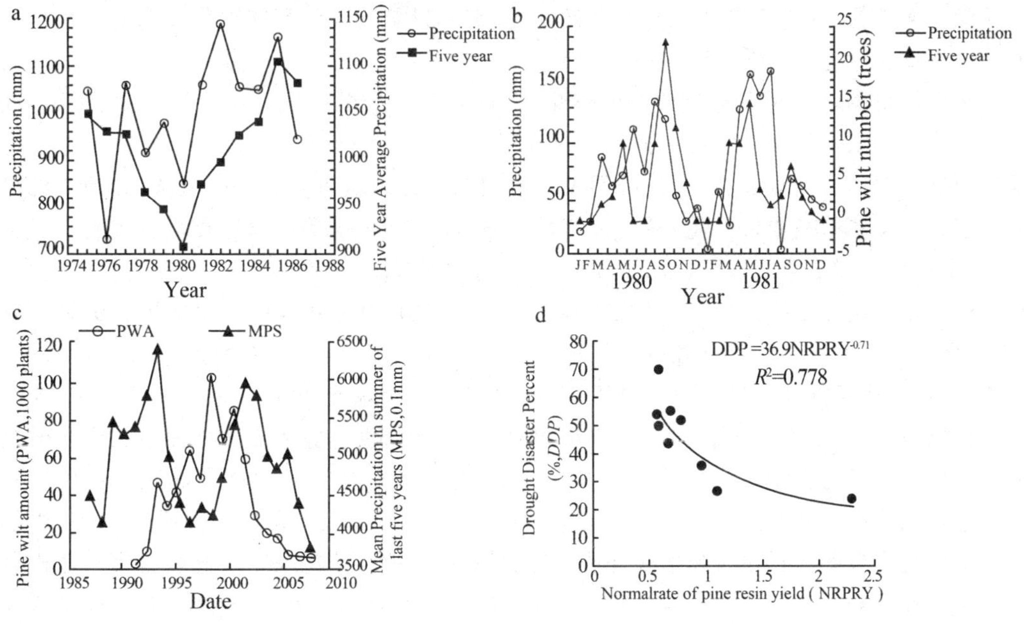

图 9　美国伊利诺斯州和中国舟山市定海区的松树枯萎情况对比。图 a 为 1974～1988 年期间伊利诺斯州降水量和前 5 年平均降水量的滑动线；图 b 为 1980～1981 年伊利诺斯州的干旱事件与松树枯萎株数；图 c 为 1985～2010 年中国浙江省舟山市定海区 5 年平均降水量和松树枯萎数量的变化趋势；图 d 为 1998 年中国松脂主产区松脂产量常年比与 1997 年干旱成灾率的相关关系。

不管怎样，这些松树枯萎病常发生的地方依然与东亚三国有相似之处，那就是它们大多分布在干热频繁且时常有台风/飓风出没的地区，尤其是极端气象事件频发的风口地带。显然，松树枯萎病的发生受所在地理环境条件的制约和影响。与东亚地区气候波动性强、台风袭击频繁的特点相比，欧美等地松树枯萎病流行的概率应该小些，尤其是几乎没有台风/飓风袭击、大部分位于北纬 40 度以北的欧洲非地中海气候区。

4. 讨论

松树对干旱环境具有独特的适应特征，再加上其树干端直和常绿习性深受人们的喜爱，故在全世界范围内被广泛引种栽培。松树的引种已经有数百年甚至上千年的历史。自然状态下松树常分布在高纬度地区和低纬度高海拔地区。在低纬度低海拔地区栽培的松树经常遭受干热和强风的胁迫。也正是由于松树具有独特的适应方式，使其在遭受强台风袭击后受害较迟，一旦受害则较重，甚至整株枯死（陈嵘，1952）。特殊的环境适应性使得松树遭受极端灾害事件袭击后常整株枯萎而少有局部的回枯，从而难于区分诱发枯萎的原生和次生作用因素。

在有充足降水的条件下，台台/飓风的机械动摇并不能对松树造成严重的伤害。几十年前就有人认为，在有充足的水分的基础上，高温和过量的光照不能对已经适应了当地环境的植物造成危害（Mansfield，Jones，1976）。在东亚的独特季风气候环境中，"出梅"之后常出现的夏季干旱时常与春秋干旱相联系，有时也与少雨台风并存。强台风、干旱、高温等极端环境因素诱发松树猝死时也大多是这些因素的集合和叠加，比如强台风之前持续的干旱少雨以及高温酷暑，台风的袭击时没有带来足以缓解干热胁迫的降水，超强的风速又使这种干热胁迫更加严重，随之而来的持续少雨天气，等等。这些极端环境超出了抗旱、耐脱水的松树的忍耐极限，导致 SPAC-连续体遭破坏直至枯死。

松树枯萎病在各个国家和地区的差异不仅与其遗传基础以及病虫危害的侵染循环有关，而且与所处的自然环境存在着密切的关联。在气候变化幅度相对较小的地中海以外的欧洲地区，由于气候环境与东亚地区有较大的差异，极端夏季干旱的袭击频率小，同时遭受台风/飓风与夏季干旱袭击的概率也相对要小，所以没有像东亚那样出现大面积的松树枯萎病的流行是完全可以理解的。在欧洲，只有在夏季炎热并常遭干热风袭击的南欧地中海地区才有松树枯萎病的研究报道（Polomski 等，2006）。在地中海沿岸国家，由于干旱事件而导致的松树枯萎也屡见不鲜（Allen 等，2010），且宏观症状与东亚流行区松材线虫病的症状相似，而不管在这些枯死松树上是否存在松材线虫。尽管澳大利亚至今尚未有松材线虫及其媒介昆虫的入侵报道（Bain，Hosking，1988），但由持续的干旱造成的松树枯萎事件在半干旱的过渡地带时有发生（Carnegie，Eldridge，2008），其外观症状类似东亚松树枯萎病流行区。澳大利亚西北部地区热带风暴和干旱事件的频繁发生与该地区引种的辐射松和海岸松的衰退（Robinson，2008）或许有内在联系。事实上，持续的严重干旱也是这些地区云杉树蜂（*Sirex noctilio*）、加利福尼亚蚜虫（*Essigella californica*）以及辐射松叶枯病（*Dothistroma septosporum*）等次生性病虫害的诱发因素（Carnegie，Cant，Eldridge，2008；Robinson，2008）。

尽管在美洲没有松树枯萎病的流行，但在一些夏季干旱和飓风袭击频繁的风口地带也有松树枯萎事件的研究报道（Dropkin 等，1981；Dropkin，Linit，1982；Malek，Appleby，1984）。1930～1935 年的持续干旱曾导致大量树木枯死，其中包括松树及其幼苗（Shirley，1934）。20 世纪 60 和 80 年代，美国马里兰州的两次干旱周期诱发了许多树木枯萎，其中始于 1962 年 5 月的干旱周期一直持续到 1967 年。1980 年 9 月持续了 2 年之久的干旱事件发生后，哈曼(Harman，1986)发现在自然分布区以外的马里兰州栽培的多脂松(*Pinus resinosa*)衰退木上存在松材线虫。其实，20世纪80年代在毗邻马里兰州的特拉华州同样发生了松树衰退的事件。哈曼发现，在没有松材线虫的枯萎植株中存在根腐病菌(*Aphelenchoidea*)和松小蠹(*Ips*)未知种，说明松材线虫只是这些多脂松的次生性病原。其实，在美国，松材线虫作为松树被害的次生性病原的报道为数不少(Harman等，1986；Bain，Hosking，1988；Wingfield，Blanchette，1982；Kozlowski，1985；Kishi，1995)。尽管飓风可造成得克萨斯州东部地区的南方松以及美国东海岸部分地区松树的机械伤害(Duryea等，2007a；Duryea等，2007b)，降低松树活力，以至于小蠹虫等虫害大量发生，甚至可以在1年之内导致数千公顷松树死亡(Pase Ⅲ，2005)，但由松材线虫病而导致的大面积松林衰退却少有报道。如前所述，日本北海道的气候特征比较接近美国东部的一些城市。与日本其他地区不同，北海道至今依然没有松材线虫病的发生和流行。1954年，超强台风15号给北海道造成了巨大损害，使铁道联络船“洞斧丸”号倾覆，也使北海道大面积森林遭到破坏，风倒木、风折木、连根拔起的树木不计其数，以至于1956～1958年小蠹虫泛滥成灾，3年间松树枯损量分别达到62.44万、61.74万和72.21万立方米(Kishi，1995)，这与经常见到的美国松林因小蠹虫爆发而受害的报道有相似之处。然而，日本北海道的气候特征不像其他县、府一样受“出梅”“入梅”的早晚及“空梅”的影响，台风与夏季干旱同时袭击的概率不大。因此，除气温相对较低外，这或许也是北海道没有发生松树枯萎病的主要原因之一。

作为多年生木本植物的松树与其周围环境构成了复杂的生态系统体系，其遭受气象等自然灾害的胁迫也是一个长期的、系统的过程，主要表现在松树枯萎与诱发事件之间的滞后现象上。也正是这种滞后为分析和判断影响因子带来了困难，并引发了一系列不必要的争议和混乱。如何透过现象发现内在的实质诱因依然是需要着力解决的问题之一，这就需要我们既注重实验研究的结果，同时更要注重大量宏观的证据及长期积累起来的数据资料(如气象资料信息等)。

主要参考文献

[1]Dropkin V. H.，Linit M. Pine Wilt—A Disease You Should Know[J]. Journal of Arboriculture，1982，8(1)：1-6.

[2]陈玉惠，叶建仁，魏初奖. 松材线虫病诊断方法进展[J]. 南京林业大学学报(自然科学版)，2001，25(6)：83-87.

[3]杨宝君，潘宏阳，汤坚，等. 松材线虫病[M]. 北京，中国林业出版社，2003，4-5.

[4]束庆龙，杨光道，邹运鼎，等. 国外松衰退病原因的研究[J]. 应用生态学报，2001，12(3)：326-330.

[5]李修鹏，王奕交. 墨西哥白松等国外松林间自然感染松材线虫病及枯死情况的调查[J]. 浙江林学院学报，1997，14(3)：273-276.

[6]熊大斌，马阅，解春霞，等. 火炬松自然条件下感染松材线虫病初报[J]. 南京林业大学学报(自然科学版)，2009，3(6)：159-161.

[7]汪来发，李占鹏，秦绪兵，等. 流胶法在长岛县防治松材线虫病中的应用[J]. 林业科学，2004，40(3)，175-178.

[8]朋金和，王晓芸，蒋丽雅，等. 安徽省松材线虫病疫情调查及检疫对策[J]. 安徽林业科技，1993，1：13-28.

[9]钱大益，沈英杰，方火良. 九江市松树枯死情况调查及原因的初步调查分析[J]. 森林病虫通讯，1991，4：12-13.

[10]李刚，黎桥华，魏小琴. 深圳市松材线虫病疫情调查[J]. 广东林业科技，1991，1：25-46.

[11]徐福元. 松材线虫病媒介昆虫的调查[J]. 森林病虫通讯，1993，2：20-21.

[12]王兴周，孙有庆，祝恩伟，等."早期诊断"在松材线虫病防治中的应用[J]. 山东林业科技，2004，4：20-21.

[13]王明旭，戴立霞，易有金，等. 湖南省松材线虫病的首次确认[J]. 中国森林病虫，2007，26(2)：5-8.

[14]黄金水，杨希，汤陈生，等. 福建省松材线虫病监测鉴定及其分布[J]. 福建林业科技，2009，36(4)：1-5.

[15]杨希，黄金水，何学友，等. 福建省马尾松枯死木中线虫检测结果分析[J]. 福建林业科技，2006，33(1)：24-27，39.

[16]黄任娥，叶建仁，潘宏阳，等. 感染松材线虫病松树滑刃目线虫调查[J]. 植物病理学报，2008，38(2)：136-146.

[17]张阜嘉. 林间松墨天牛的诱捕及其所携带松材线虫的快速检测和疑似病木的早期诊断[M]. 厦门：厦门大学出版社，20-22.

[18]Kishi Y. Forest Pest No. 1：The Pine Wood Nematode and the Japanese Pine Sawyer[M]. Tokyo，Japan：Thomas Company Limited，1995，43-48.

[19]Wingfield M. J. Transmission of Pine wood Nematode to Cut Timber and Girdled Trees[J]. Plant Disease，1983，67：35-37.

[20]迟树友，韩正敏，何月秋. 无菌松材线虫对10年生黑松致病性的研究[J]. 林业科学，2006，42(10)：71-73.

[21]Japan National Forest Pest Society(JNFPS). Recent Research and History of Japanese Pine Sawyer (pine wilt disease)[M]. Kyobun Ltd，1997，1-43. (In Japanese)

[22]徐克勤，徐元福. 干旱胁迫对松材线虫病的影响[J] 南京林业大学学报，1996，20(2)：80-83.

[23] Rutherford T. A.，Webster J. M. Distribution ofPine Wilt Disease with Respect to Temperature in North America，Japan，and Europe[J]. Canadian Journal of Forest Research，1987，17：1050-1059.

[24]Mamiya Y. The Pine Wood Nematode. In：Nickle，. W. R. (Ed.). Plant and Insect Nematodes[M]. New York：Marcel Dekker，1984：589-626.

[25]Suzuki K. General Effect of Water Stress on the Development of Pine Wilt Disease Caused by Bursaphelenchus Xylophilus[J]. Bull. For. & For. Prod. Inst.，1984. 325，97-126.

[26]J. D. Jarrell，P. J. Hebert，B. M. Mayfield. Hurricane Experience Levels of Coastal County Populations—Texas to Maine[J]. NOAA，Technical Memorandum NWS-NHC-46，1992，152.

[27]陈嵘. 造林学特论[M]. 南京：中国图书发行公司南京分公司，1952，136-139.

[28] Mansfield T. A.，Jones M. B. Photosynthesis：Leaf and Whole Plant Aspects. In Hall M. A. eds. Plant Structure，Function and Adaptation[M]. Landon，Basingstroke：Macmillan Press LTD.，1976，315-316.

[29]J. Polomski，U. Schonfeld，H. Braasch，et al. Occurrence of Bursaphelenchus Species in Declining Pinus Sylvestris in Dry Alpine Valley in Switzerland[J]. Forest Pathol，2006，36：110-118.

[30]Allen C. D.，Macalady A. K.，Chenchouni H.，et al. A Global Overview of Drought and Heat-induced Tree Mortality Reveals Emerging Climate Change Risks for Forests[J]. For. Eco. Manage，2010，259(4)：660-684.

[31]Bain J.，Hosking G. P. Are NZ Pinm Rradiata Plantations Threatened by Fine Wilt Nematode Bwsaphelenchm Xylophilus? [J]. New Zealand Forestry，1988，32(4)：19-21.

[32]Carnegie A. J.，Cant R. G.，Eldridge R. H. Forest Health Surveillance in New South Wales，Australia [J]. Australian Forestry，2008，71 (3)：164-176.

[33]Robinson R. Forest Health Surveillance in Western Australia：A Summary of Major Activities from 1997 to 2006[J]. Australian Forestry，2008，71(3)：202-211.

[34]Dropkin V. H.，Foudln A.，Kondo E.，et al. Pine Wood Nematode：A Threat to U. S. forest? [J]. Plant Disease，1981，65(12)：1022-1027.

[35]Malek R. B.，Appleby J. E. Epidemiology of Pine Wilt in Illinois[J]. Plant Disease，1984，68(3)：

180-186.

[36] Shirley H. L. Observations onDrought Injuries in Minnesota forests[J]. Ecology, 1934, 15(1): 42-48.

[37]Harman A. L., Krusberg L. R., Nickle W. R. Pinewood Nematode, Bursaphelenchus Xylophilus, Associated with Red Pine, Pinus Resinosa, in Western Maryland[J]. Journal of Nematology, 1986, 18(4): 575-580.

[38] Wingfield M. J. ,Blanchette R. A. Association of Pine Wood Nemantode with Stressed Trees in Minnedota, Iowa, and Wisconsin[J]. Plant Disease, 1982, 66: 934-937.

[39]Kozlowski T. T. Tree Growth in Response to Environmental Stresses[J]. J. Aboriculture, 1985, 11(4), 97-111.

[40]Duryea M. L., Kampf E., Littell R. C. Hurricanes and the Urban Forest: Ⅰ. Effects on Southeastern United States Coastal Plain Tree Species[J]. Arboriculture & Urban Forestry, 2007a, 33(2): 83-97.

[41]Duryea M. L., Kampf E., Littell R. C., et al. Hurricanes and the Urban Forest: Ⅱ. Effects on Tropical and Subtropical Tree Species[J]. Arboriculture & Urban Forestry, 2007b, 33(2): 98-112.

[42]Pase Ⅲ(Jeo) H. A. Evaluation and Management of Storm-Damaged Timber[J]. Pest, 2005, 10(3): 1-2.

(译自 *Natural Hazards*,2014,6,72: 723-741)

第六部分　其他

光热和水分条件对石灰岩山地侧柏人工林更新的影响*

王　斐[1]　臧丽鹏[2]

(1. 山东省林业科学研究院，济南，250014；2. 中国林业科学研究院，北京)

侧柏(*Platycladus orientalis* (L.) Franco)系单种属常绿鳞叶树种。侧柏天然林主要分布于中国华北以及东北、华东、西南局部山区，人工栽培遍及中国各地，尤其是干旱瘠薄的石灰岩山地。侧柏对土壤的适应性广，从石灰性土壤到棕壤、褐土等均可生长，尤其是在其他树种难以生存的裸岩山地，常可见到丛生的灌木状侧柏林。在这种干旱瘠薄的土壤上，侧柏群落相对稳定，其耐干旱瘠薄的特性也赢得了人们的青睐，以至于成了我国华北、西北等地区主要的荒山造林树种[1]。尽管侧柏结实量很大，种子易于萌发，但在侧柏人工林的干旱空旷阳坡环境和密闭的阴坡环境下其苗木和幼树并不多见。为此，我们在山东省侧柏主要栽培区济南、淄博、枣庄等地进行了详细的考察。结果发现，侧柏幼苗常集中生长于母树树冠侧下方，伴随着幼苗的成长势必出现严酷的种内竞争。其中一些苗木能成功地进入树冠层，而另一些则严重受压而弯曲、衰退甚至枯萎。除林缘和林窗以外，在郁闭的树冠下面难以见到新生幼苗的踪迹。尽管国内对侧柏的研究较多[2~6]，但是针对侧柏林窗幼苗的空间分布及幼苗适应环境特殊性的研究并不多见。本研究通过热红外成像法近距离移动观测了山体光热能量分配的地面真实特征，用 Sketch UP 软件进行了计算机模拟林窗光照[7]的分布情况，追踪了林窗阴影的踪迹。在此基础上，发现了侧柏幼苗分布和生长与林窗光热能量再分配有关，揭示了侧柏幼苗对环境具有特殊适应性。类似的研究有可能成为促进侧柏生态公益林天然更新，提高侧柏林地可持续性山地防护持能力的重要依据。

* 基金项目：国家自然科学基金(31170671)；山东省科学研究计划项目(2012GNC1117)。第一作者简介：王斐(1959～)，男，山东临邑人，日本鸟取大学农学博士，研究员；主要从事林木栽培管理、气象学、树木生理生态学以及光谱分析等研究。

1. 材料与方法

1.1 红外热像温度的移动观测

在山东省，除了少量的零星栽培以外，侧柏大多栽培在山地环境中，尤其是石灰岩山地。而山地环境与平原的最大区别在于其光能在不同坡向和坡位的再分配[8,9]。山地环境的土壤湿度也不尽相同[8]。热红外成像技术不仅可以反映观测对象的温度，而且在一定程度上也提供了其湿润状况（含水率等）的信息[10]。因此，本研究设计了山林地近距离热红外图像的地面真实观测研究。红外热像用 NEC H2640 型热红外相机（波长 8～13 μm）拍摄，测温范围为－40～500 ℃，最小感温能力 0.03 ℃。在设定红外像机为自动灵敏度和 0.98 的发射系数的状态下，手持热红外摄像机于目标林地上方 150 cm 处调焦清晰后拍摄红外热像。

鉴于热红外成像设备价格不菲，因此使用多台同型号的设备同时在多点拍摄不太现实。为了尽可能同时观测同一山体不同方位的热象温度值以增加数据的可比性，我们设计了应用一台热红外相机移动拍摄红外热像的观测方法[11]，定时观测了济南市馍馍山（东经 117°03′28″，北纬 36°08′13″）8 个方位的热象温度。馍馍山山体浑圆，是进行山地光照和热能分配特征观测的理想环境。热象是于山体中部侧柏林木比较均匀、树冠完全遮蔽且没有地被物的地面上拍摄的。每个观测点重复拍摄 5 次。观测点的方位分别是北、东北、东、东南、南、西南、西、西北，从早 9 点到下午 3 点，每小时一个轮回。结果按公式(1)计算各点的平均热温数值：

$$IT_i = \sum_{j=1}^{6} t_{ij} \tag{1}$$

式中，IT_i是第 i 个方位的平均热象温度，$i=1,2,\cdots,8$，t_{ij}是第 i 个方位在 j 时的热象温度值，$j=1,2,\cdots,6$。

1.2 林地指温差指数和土壤含水量的对应观测

2015 年夏季，在山东省林业科学研究院燕子山林场（东经 117°03′36″，北纬 36°38′58″）侧柏生态公益林的阳坡（东南、南和西南）和西坡（西和西北），于太阳落山 1 小时后，以观测者手指的热像温度为参照，应用 NEC H2640 型热红外相机在不同密度的疏伐林地（扒开枯枝落叶层）的土壤上方 30～50 cm 处拍摄红外热像；然后应用公式(2)计算土壤的指温差指数[12]：

$$TD_{lf} = \sum_{i=1}^{n} (Tf_i - Tl_i) \tag{2}$$

式中，TD_{lf}是指温差指数，Tl_i为第 i 个重复观测的热温数值，Tf_i为第 i 个重复观测的手指热温数值，n 为重复数，一般重复 5 次。

在拍摄热红外图像的同时采取土样，用称重法测定其含水率，并进行对比分析。

1.3 山地侧柏林窗日照率和日照时数分布特征的 Sketch UP 软件模拟

Sketch Up 是一种简便、易用的 3D 设计软件，可按物体地理空间位置和时间序列显示其阴影，适合实时解析树阴和日照的关系。洪文祥(2012)[13]曾应用该软件研究了高层建筑物阴影的观测精度及阴影对花木的影响。安洁玉等(2010)[14]用此软件配合谷歌地球(Google Earth)的二维影像计算了建筑物的高度等。本研究以山东省林业科学研究院燕子山林场侧柏生态公益林为模板，应用 Google Sketch UP 2014 软件建立了密度 3000 株/公顷、树高 6～7 m、冠幅 2.1 m 的模拟侧柏林地，并分别在东南阳坡和西北阴坡(坡度 30°)设计出长约 10 m、宽 5～6 m 的数字化模拟现实林窗。从早上 5 点到晚上 19 点 30 分，以 30 分钟为时间间隔，分别截取每个林窗夏至日的 RGB 图像，然后用点阵法获取林窗内光照或阴影的分布信息。依据林窗大小，分别在东南坡和西北坡的林窗内设置 146 和 197 个样点，然后将阴影样点数量化为 1，光亮样点为 0，经统计计算每幅图像的阴影分布数据，绘制日照时数分布图和林窗日照率的日变化曲线图，以此研究了林窗光照的时空轨迹。其中，林窗日照面积率的数值计算见公式(3)：

$$IR_{\mathrm{i}} = \frac{100 \times \sum_{j=1}^{m} X_{ij}}{\left(\sum_{k=1}^{n} Y_{ik} + \sum_{j=1}^{m} X_{ij}\right)} \tag{3}$$

式中，IR_i 为第 i 时林窗日照面积率，X_{ij} 为第 i 时第 j 个赋值为 0 的样点，$j=1,2,\cdots,m$；Y_{ik} 为第 i 时第 k 个赋值为 1 的样点，$k=1,2,\cdots,n$。

林窗内各样点的日照时数值的计算见公式(4)：

$$IP_{i} = \sum_{j=1}^{m} X_{ij} \tag{4}$$

式中，IP_i 为第 i 个林窗样点的日照时数(以半小时为一个单位)，X_{ij} 为第 i 个样点在第 j 时的赋值为 0，$j=1,2,\cdots,m$。

为了客观地反映疏伐林地内侧柏幼苗和幼树上方的光照状态，我们应用 CI110 型冠层分析仪观测了侧柏幼苗和幼树上方的透光度数值。在操作方法上，为了尽可能地避免强光的影响，观测基本上是在光线比较柔和的条件下进行的。观测时，使用设备的默认设置：视域范围为 150，初始天顶角 0°，结束天顶角 70°，阈值 50。

1.4 侧柏林窗更新苗数和苗高的观测

为了研究侧柏幼苗和幼树在林窗内的发生和分布受光照和遮阴条件的影响，分别以山地的东南、南和西北坡作为观测地点，选择典型的林窗各 5 个，大小一般约 5 m×10 m。依据林窗的方位将林窗分为遮阴面和透亮面。然后分别计数各面的幼苗和幼树棵树，同时用钢卷尺测量每株苗木/幼树的高度。为了增加可比性，避免新生苗数量的影响，对苗高进行统计分析时常用最高的 50～170 株幼苗/幼树的高度作为基本统计数据。

1.5 数据处理

对更新苗木调查数据进行 Fisher 分布统计假设检验，并将检验结果的 F 值置于图内相应的位置，其中“ * ”是指 95％的概率差异显著，“ * * ”是指 99％的概率差异显著。统计计算用 Microsoft Excel 2003 软件完成，相应的制图用 Google Sketch UP 2014、Microsoft Excel 2003 和 Photoshop CS 等软件完成。

2. 结果与分析

2.1 山地和林窗环境光热能量的再分配、土壤湿度和播种出苗率

本研究经过精心设计，进行了山地热红外成像的移动观测。经侧柏山林地 8 个方位(北、东北、东、东南、南、西南、西、西北)6 小时内的巡回观测，其平均热温值如图 1a 所示。显然，位于阳坡的南、东南和西南各坡向的光照和热量明显高于北、东北和西北的阴坡环境，而半阴半阳的东和西坡介于其中。阳坡的正南和西南方位的热温最高，北和西北坡接受热能少、热温最低，相对阴湿。应用 Sketch UP 软件模拟山地侧柏林窗内的光照特征表明，不同坡向差异较大。以东南阳坡和西北阴坡为例，东南坡主要在下午 2 点之前接受阳光照射，而光照高峰时段出现在 10～12 点之间，这时太阳高度角大，在正午阳光的照射下地面接收能量多而强(见图 1b-东南坡)。西北坡主要在上午 11 点以后接受阳光照射，光照高峰时段出现在 14～16 点之间，这时太阳高度角渐小，斜射的阳光受树冠的影响使地面接收能量较少且弱(见图 1b-西北坡)。另一方面，无论是东南坡还是西北坡，林窗的北面均接受相对较多的光照。东南坡的西北向和西北坡的东北向往往是日照时数最高的地段。无论东南还是西北坡向，林窗内均可区分出明显的遮阴面(日照时数 0～7 小时)和透光面(日照时数 8～14 小时)(见图 1c、1d)。

2015 年夏，在不同疏伐强度(每 666.7 平方米保留株树分别为 80、105、120、135、170、200)和坡向的林分内观测到了土壤含水率与指温差指数的正相关关系(见图 2a)。这意味着在疏伐强度较大的林地内，林地透光度大，光照相对充足；土壤热温较高，指温差指数也较小，土壤蒸发速率相对较快，土壤含水率相对较低。相反，在疏伐强度较小或者未经疏伐的林地内，林地透光度小，光照相对不足，土壤热温较低，指温差指数也越大，土壤蒸发速率相对较慢，土壤含水率相对较高。如图 1a 所示，这种相关关系也受坡向的影响：阳坡接收的太阳辐射较多，地温偏高，土壤相对干燥；阴坡接收的太阳辐射较少，地温偏低，土壤相对湿润。

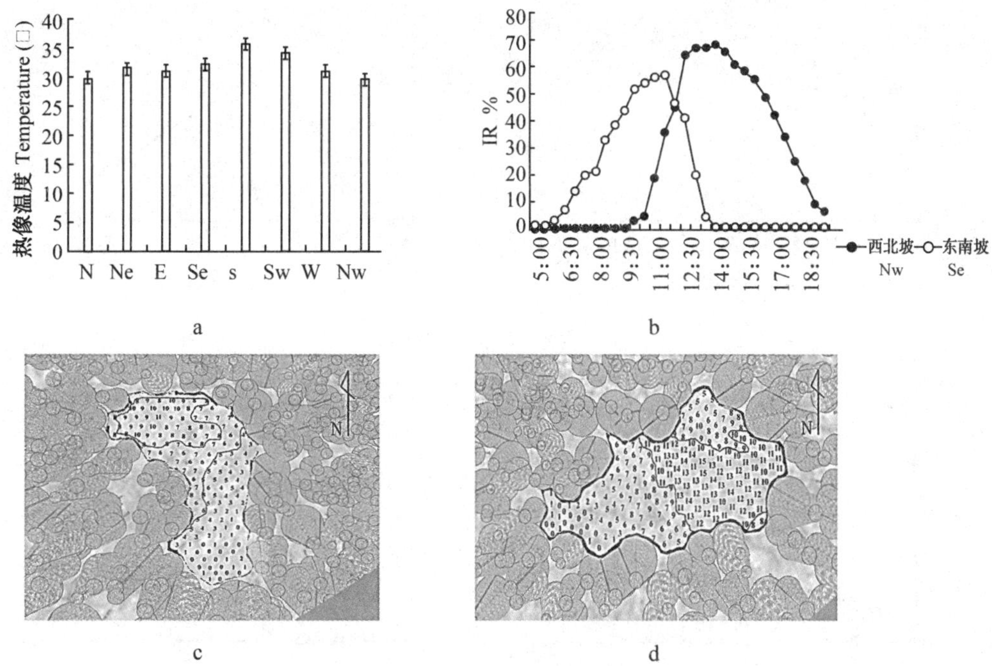

图 1　山地侧柏林和林窗光和热量分布的移动观测和计算机模拟：图 a 为济南林场馍馍山侧柏林不同方位（其中 N-北，Ne-东，E-东，Se-东南，N-南，Sw-西南，W-西和 Nw-西北坡）的热象温度移动观测值；图 b 为 Sketch UP 软件模拟山地不同方位的侧柏林窗光照面积率（IR）的日变化趋势；图 c 为 Sketch UP 软件模拟山地东南坡侧柏林窗内各样点日照时数（IP）的空间分布特征；图 d 为 Sketch UP 软件模拟山地西北坡侧柏林窗光照时数（IP）的空间分布特征。

在山东济南，经历了 2014 年度的少雨干旱之后，2015 年夏季依然降水不足。因此，在上述疏伐林地中播种充分催芽的种子，其出苗率与林地透光度之间呈反函数相关关系（见图 2b），与林地的指温差指数之间呈显著的正相关关系（见图 2c）。这说明，综合反映林地光照和湿度的指温差指数比仅反映光照状态的透光度与出苗率的关系更密切，对出苗率的影响也更大。显然，在相对干旱的环境中侧柏种子的萌发受土壤湿度的影响较大，在土壤相对干旱、指温差指数较小的阳坡播种出苗率明显偏低。而在相对阴湿、指温差指数较大的阴坡播种出苗率明显较高。相对阴湿的未疏伐林地中出苗率高于强度疏伐的林地。然而，由于试验林地密度较大，未疏伐的地块每 666.7 平方米中有侧柏 200 株以上，林地光照不足，籽苗地上、地下生长不平衡；雨季过后，侧柏籽苗很快因根系发育不良而干枯。所以，侧柏籽苗保存率与林地透光度之间呈明显的正相关关系（见图 2d）。

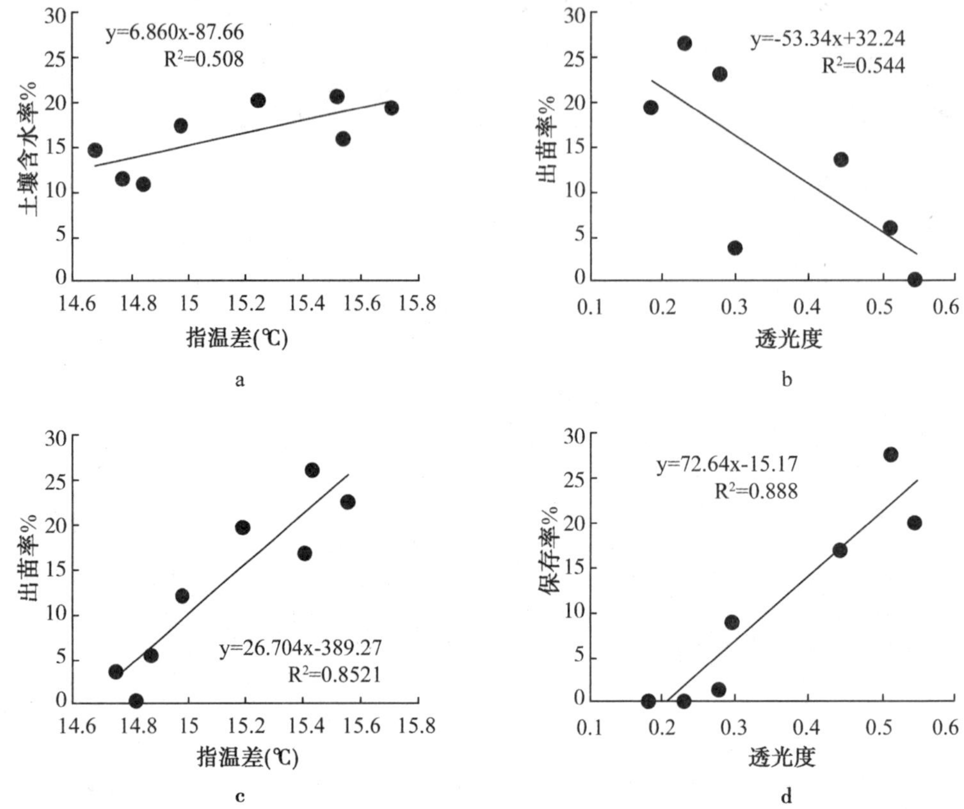

图 2　侧柏播种出苗率/保存率、土壤含水率与指温差指数和林地透光度的关系：图 a 为土壤含水率与指温差指数的相关关系；图 b 为出苗率与透光度的相关关系；图 c 为出苗率与指温差指数的相关关系；图 d 为保存率与指温差指数的相关关系。

2.2　山体异向林窗内侧柏幼苗的分布

在侧柏生态公益林中，中幼林的林分常因密度过大而形成单层纯林。在这些过密的林分下，很少能看到侧柏幼苗和幼树。侧柏苗大多分布在林窗、林缘附近。在接受阳光最多的东南、南和西南阳坡上，林窗内苗木大多分布在侧方庇荫的一面（见图 3a）；在侧方遮阴的环境条件下，侧柏幼苗不仅数量多，而且长势好、苗高也高（见图 3a）。在接受阳光较少的北和西北坡，幼苗和幼树主要分布在透亮的一面，遮阴面苗木数量少、长势差、苗木低矮（见图 2b）。个别情况下，遮阴面甚至没有幼苗存在（见图 3b）。

同样是西北坡，尽管遮阴面幼苗数量和质量均不如透光的一面，下坡位往往幼苗数量较少，长势弱（见图 2c），而中上坡位幼苗数量较多，长势旺盛（见图 3d）。中上坡位与林窗透亮区域的相似之处在于二者均接受更多的光照。显然，光热能量在不同坡向和坡位上的再分配直接导致了侧柏的生长和更新的差异。侧柏幼苗在干热的环境中喜欢一定的遮阴，而在阴湿的环境中又需要足够的光照来维持生存。也就是说，侧柏幼苗需要在优化的光照和水分条件下才能正常生长。在阳坡光照相对充足的环境中，过分干燥是一大限制因素，幼苗和幼树常出现在侧方庇荫的一面。而在阴坡光照相对匮乏的环境中，光能不足是限制性因素，侧柏幼苗和幼树大多生长在透光的一面。

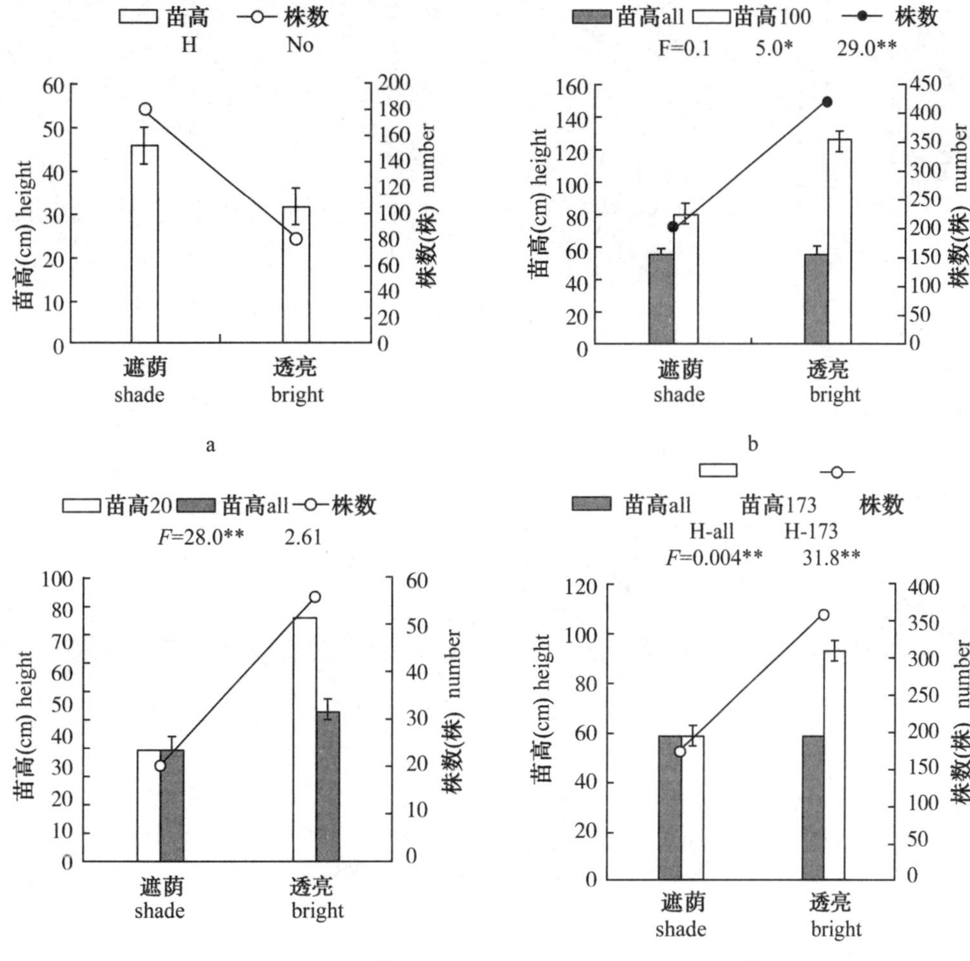

图 3 侧柏更新苗木数量和高度实测结果：图 a 为阴坡（西北坡），图 b 为阳坡（西南坡），图 c 为下坡位（西北坡），图 d 为中上坡位（西北坡）。其中，苗高$_{all}$意指全部苗木的平均高；苗高$_{173}$意指最高的 173 株优势苗木平均高度，以此类推。

2.3 特殊立地条件下侧柏幼苗的更新和生长

进一步深入调查发现，在一些干旱的阳坡坡面上，即使林木密度和光照条件适中，侧柏幼苗和幼树也很少（见图 4a-正常）。林地中局部汇水面常密集生长着众多幼苗和幼树（见图 4a-汇水坡面）。在山地水沟边缘同样着生较多的幼苗，由于土壤水分有效性的差异，同一条水沟的上部（见图 4b）往往比下部的幼苗数量少，生长量也小（见图 4b）。

在相对比较干旱、林地尚未郁闭的环境中，同一山体的西北坡接受太阳光能少，相对阴湿，侧柏幼苗、幼树数量较多（见图 4c），个体大（见图 4d）。由于山体的东南坡接受太阳光能相对较多、相对干燥，因此侧柏幼苗、幼树数量少（见图 4c），个体也小（见图 4d）。

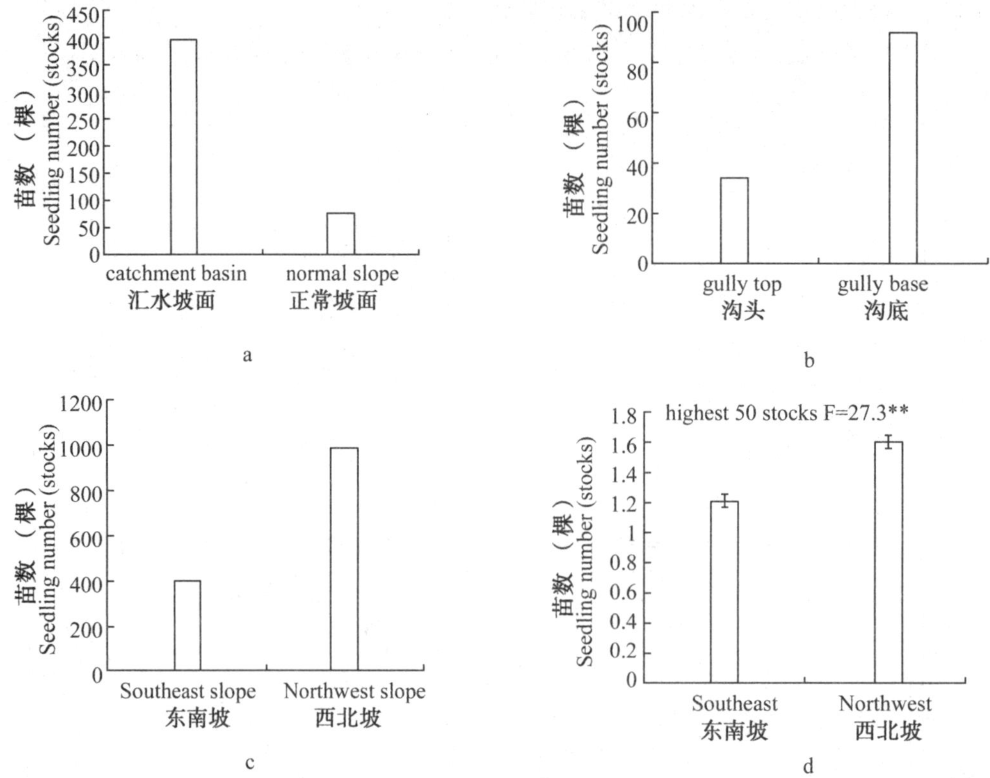

图 4　干瘠山地环境中侧柏天然更新苗分布特征：图 a 为汇水坡面与正常坡面立地的更新苗木数量；图 b 为沟头和沟底的更新苗数量；图 c 为东南坡与西北坡的更新苗数；图 d 为东南坡与西北坡苗高（上层 50 株）的差异。

3. 讨论

本研究通过热红外成像技术和 Sketch UP 软件的计算机林窗阴影模拟，用不同方法取得了与山地气象观测研究相似的结果[8]，且与近年来有关林窗光照环境的研究结果相吻合[7]。相比之下，热红外成像技术和计算机林窗阴影模拟更加方便快捷，为研究森林小气候、森林更新等开辟了新径。

目前人们普遍认为侧柏是喜光树种，同时侧柏幼林能耐受一定程度的庇荫[1]。本研究的结果在一定程度上支持这一论点，但与此论点并不完全相同。树木的耐阴性通常是指它们在林冠下生活的能力[16]。就郁闭的侧柏林下少有苗木和幼树而论，侧柏的确是喜光树种。在密度较大的侧柏林分中，幼苗和幼树往往因严重受压而弯曲、衰退甚至枯萎。一些植株向光生长而使主干弯曲，个别植株几乎是从直立直接横向生长。另一方面，树木在林下不仅受光照环境的影响，还受土壤湿度[16]等因素的限制。在北和西北坡相对湿润的环境中，干热等并非限制性因素，侧柏苗木多生长在透光的一侧；在庇荫的一侧或者没有幼苗和幼树，或者生长较弱，这似乎与幼苗耐阴产生了矛盾。在光线不足的环境中难以生存也进一步证明侧柏是喜光的(包括幼树和幼苗)。然而,在南或东南阳坡相对干燥的环境中,光照并非限制性因素[16]。土壤湿度等成了侧柏苗木长存的限制因素,侧柏苗往往只生长在侧方或上方庇荫的条件下,而在无庇荫的全光照环境下则几乎没有侧柏幼苗的存在。就此而论,侧柏幼苗并非是"耐阴"的,而应理解为是"喜湿"的。也就是说,侧柏是喜光树种,幼时稍不耐旱,尤其是在阳坡林窗内的干热区域。

研究发现,在一些干旱瘠薄的阳坡甚至在全光照环境条件下,侧柏幼苗大量生长于山地汇水面、水沟两侧以及山地沟谷两边,这说明侧柏幼时喜光、喜湿润环境,这似乎与目前普遍接受的侧柏"耐干旱瘠薄"的观点相矛盾。其实,侧柏正是在这种矛盾的过程中成长的。侧柏之所以耐干旱瘠薄,除了其鳞叶保水能力强外,似乎也与适应性强的根系有关。据报道,侧柏的根系随环境的不同而表现出很大的可塑性[17]:侧柏的水平根发达,时而也有斜生主根。细根大多分部在表层 10~30 cm 的土壤范围内,而且其在土层较厚的环境中能伸展得更深[2,5,17]。侧柏基本上属浅根系树种[5,6],横向生长的水平根更加适合在薄层的山地土壤中生长,在土壤水分匮乏时有足够的能力横向寻找水源。有报道称,侧柏主根在风化的花岗岩层中可沿岩缝向下伸延 1 m 多[17]。侧柏幼苗和幼树之所以不耐干旱,这是因为侧柏幼苗根系尚未完善。据观察,侧柏幼苗的根系呈主根系状,水平根和侧根并不发达,短时间内很难将根系拓展到有足够"水源"的地方。而在侧柏母树庇荫下的相对阴湿环境中对水分的消耗更少,故比较适合侧柏的生长。

主要参考文献

[1]许牧农,侧柏林,徐化成等主编. 中国森林[M]. 北京:中国林业出版社,1999.

[2]周长瑞,孙怀锦,郭善基. 论侧柏[J]. 山东林业科技,1981(3):63-67.

[3]程高芳,张志诚. 太行山区侧柏天然更新规律初探[J]. 河南林业科技,1988(1):37-38.

[4]鲁法典,韩峻,李超. 泰山侧柏林下幼树分布规律及天然更新研究[J]. 河北林果研究,2006,21(1):29-31.

[5]刘元本等. 河南森林[M]. 北京:中国林业出版社,2000.

[6] 冯林等. 内蒙古森林[M]. 北京:中国林业出版社,1989.

[7]王进欣,王今殊,张一平等. 西双版纳干季人工林林窗边缘增温效应初步分析[J]. 徐州师范大学学报(自然科学版),2002,20(2):61-65.

[8]傅抱璞. 山地气候[M]. 北京:科学出版社,1983.

[9]赵胜亭,宋秀英. 基于 DEM 的栖霞山区日照时数模拟研究[J]. 农业网络信息,2011(5):15-18.

[10]Wang F., Yamamoto H., Ibaraki Y., et al. Detecting Leaf Necrosis and Branch Dieback of Kousa Dogwood Trees after Transplanting Shock by Amplified Variation of Imaging Temperature[J]. Journal of Agricultural Meteorology, 2010, 66(2): 81-90.

[11]Takayama N., Yamamoto H., Zhang J. Q., et al. Development and Application of Mobile Observation Method of Urban Climate To Analyze Occurrence of Heat-Island in Changchun City of China[J]. AIJ Journal of Technology and Design, 2009, 15(31): 827-832. (In Japanese)

[12]王斐,吴德军,翟国锋,等. 侧柏衰弱木和蛀干害虫受害木的热红外成像检测[J]. 光谱学与光谱分析,2015,35(12):3410-3415.

[13]洪文祥. 武汉市建筑阴影分析及绿化植物选择的研究[D]. 华中农业大学,2012.

[14]安洁玉,程朋根,丁斌芬. 基于 Google Earth 二维影像获取建筑物高度的方法[J]. 地理与地理信息科学,2010,26(6):31-33,37.

[15] Wang F. ,Omasa K. Image Measurements of Leaf Scorches on Landscape Trees Subjected to Extreme Meteorological Event [J]. Ecological Informatics, 2012, 12: 16-22.

[16]斯波尔、巴恩斯著,赵克绳、周祉译. 森林生态学[M]. 北京: 中国林业出版社, 1981.

[17]向师庆. 北京主要造林树种的根系研究[J]. 北京林学院学报, 1981(2): 19-32.

(原文发表于《防护林科技》,2016 (7):1-6.)

致　谢

本文集中论文的发表和相关研究得到了国家自然科学基金项目“用热红外成像技术研究一些树种对极端干热环境的响应特征和机理”“数字、热红外和显微图像解析法对一些树木水分和能量失衡的解析”和山东省科学技术计划研究项目“侧柏生态公益林抚育经营关键技术研究”“侧柏林更新与能源综合利用的研究”的资助，在此特表示衷心的感谢！